Complex Adaptive Systems

Complex Adaptive Systems

AN INTRODUCTION TO COMPUTATIONAL MODELS OF SOCIAL LIFE

John H. Miller and Scott E. Page

PRINCETON UNIVERSITY PRESS

PRINCETON AND OXFORD

Copyright © 2007 by John H. Miller and Scott E. Page

Requests for permission to reproduce material from this work should be sent to
Permissions, Princeton University Press

Published by Princeton University Press, 41 William Street,
Princeton, New Jersey 08540

In the United Kingdom: Princeton University Press, 3 Market Place,
Woodstock, Oxfordshire OX20 1SY

All Rights Reserved

Library of Congress Control Number: 2006933230

ISBN-13: 978-0-691-13096-5 (acid-free paper)
ISBN-10: 0-691-13096-5 (acid-free paper)

ISBN-13: 978-0-691-12702-6 (pbk.: acid-free paper)
ISBN-10: 0-691-12702-6 (pbk.: acid-free paper)

British Library Cataloging-in-Publication Data is available

This book has been composed in Sabon

Printed on acid-free paper. ∞

press.princeton.edu

Printed in the United States of America

7   9   10   8   6

To George Cowan

Contents

Figures

Tables

Preface

> We have to look for routes of power our teachers never
> imagined, or were encouraged to avoid.
> —*Thomas Pynchon, Gravity's Rainbow*

THE EMERGING TAPESTRY of complex systems research is being formed by localized individual efforts that are becoming subsumed as part of a greater pattern that holds a beauty and coherence that belies the lack of an omniscient designer. As in Navajo weaving, efforts on one area of this tapestry are beginning to meld into one another, leaving only faint "lazy lines" to mark the event. The ideas presented in this book contain various parts of this weaving; some are relatively complete, whereas others are creative investigations that may need to be removed from the warp and started anew. We suspect that, like the Navajo weavers of old, we will also introduce a few errors—though perhaps not intentionally—that will be more than sufficient to maintain our humility.

More than a decade ago, a wonderful coincidence of people, ideas, tools, and scientific entrepreneurship converged at the Santa Fe Institute. Those of us who participated in this event were blessed to partake in a burst of scientific creativity that facilitated a new wave in the sciences of complex systems. At that time, discussions about the central problems and approaches in fields such as biology, chemistry, computer science, economics, and physics made it clear that there was a common set of questions that would require a willingness to transcend the usual disciplinary boundaries if answers were to be forthcoming. Since that time, a growing community of scholars has been actively involved in developing the theory of complex adaptive social systems.

Although research in the area of complex adaptive social systems is still in its formative stages, now is a good time to take stock of these efforts. Along with documenting much of what we have learned over the past decade, we will also be a bit exploratory, both retrospectively trying to figure out why our initial intuitions about the importance of this area were justified and prospectively suggesting where the new frontiers are likely to be found.

During the past decade we have hosted an annual graduate workshop in computational modeling. In these workshops, we collaborated with a diverse set of graduate students who are interested in applying new computational modeling techniques to key problems in the social sciences. Many of the topics presented throughout this book are the result of discussions during these workshops.

Contrary to the sentiments in Pynchon's quotation, we have been blessed with some very imaginative and prescient teachers. For Miller, Ken Boulding planted the initial meme that suggested that both biological and social systems hold a deep similarity needing scientific investigation. Ted Bergstrom and Hal Varian generously indulged and guided Miller's efforts during graduate school in investigating the behavior of artificial adaptive agents in games. Bob Axelrod, John Holland, and Carl Simon were also sources of encouragement, ideas, and wisdom at that time. During the early days of the Santa Fe Institute, an outstanding group of scholars gathered together to work on complex systems, including Phil Anderson, Ken Arrow, Brian Arthur, George Cowan, Jim Crutchfield, Doyne Farmer, Walter Fontana, Murray Gell-Mann, Erica Jen, Stu Kauffman, David Lane, Blake LeBaron, Norman Packard, Richard Palmer, John Rust, and Peter Stadler, all of whom have contributed in various ways to the ideas presented here. Miller's colleagues at Carnegie Mellon University, in particular Greg Adams, Wes Cohen, Robyn Dawes, George Loewenstein, John Patty, and especially Steven Klepper, have been a continual source of ideas and encouragement, as has been Herb Simon, whose contributions to complex systems and social science will continue to inspire and craft research efforts far into the future.

For Page, his graduate adviser Stan Reiter organized a group of students to investigate research on learning, adaptation, and communication, and these discussions eventually led him to the Santa Fe Institute to learn more about complex systems. At that time, a lively and ongoing collaboration that focused on computational political economy was started among the authors and Ken Kollman. While at the California Institute of Technology, Page benefited from many discussions about mathematics, theory, complexity, and experiments, with Mike Alvarez, John Ledyard, Richard McKelvey, Charlie Plott, and Simon Wilkie. Page's current colleagues in the Center for the Study of Complex Systems at the University of Michigan, including Bob Axelrod, Jenna Bednar, Dan Brown, Michael Cohen, Jerry Davis, John Holland, Mark Newman, Mercedes Pascual, Rick Riolo, Carl Simon, and Michael Wellman, as well as his collaborator Lu Hong, have also been extremely influential.

The authors wish to thank various students and seminar participants across the world who have been kind enough to give us additional insights into these ideas. In particular, Aaron Bramson, Scott deMarchi, and Jonathan Lafky provided some detailed input. Chuck Myers at Princeton University Press has also provided wonderful encouragement and direction, and Brian MacDonald thoughtfully copyedited the manuscript.

Some of the nicest examples of interesting complex social systems have emerged in our home institutions. We are grateful to the research infrastructure of the Santa Fe Institute, Carnegie Mellon University, and the University of Michigan. In particular, we would like to thank Susan Ballati, Ronda Butler-Villa, Bob Eisenstein, Ellen Goldberg, Ginny Greninger, George Gumerman, Ginger Richardson, Andi Sutherland, Della Ulibarri, Laura Ware, Geoffrey West, and Chris Wood at the Santa Fe Institute; Michele Colon, Carole Deaunovich, Amy Patterson, Rosa Stipanovic, and Julie Wade at Carnegie Mellon University; and Mita Gibson and Howard Oishi at the University of Michigan.

PART I

Introduction

CHAPTER 1

Introduction

> The goal of science is to make the wonderful and complex
> understandable and simple—but not less wonderful.
> —*Herb Simon, Sciences of the Artificial*

> The process of scientific discovery is, in effect, a continual
> flight from wonder.
> —*Albert Einstein, Autobiographical Notes*

ADAPTIVE SOCIAL SYSTEMS are composed of interacting, thoughtful (but perhaps not brilliant) agents. It would be difficult to date the exact moment that such systems first arose on our planet—perhaps it was when early single-celled organisms began to compete with one another for resources or, more likely, much earlier when chemical interactions in the primordial soup began to self-replicate. Once these adaptive social systems emerged, the planet underwent a dramatic change where, as Charles Darwin noted, "from so simple a beginning endless forms most beautiful and most wonderful have been, and are being, evolved." Indeed, we find ourselves at the beginning of a new millennium being not only continually surprised, delighted, and confounded by the unfolding of social systems with which we are well acquainted, but also in the enviable position of creating and crafting novel adaptive social systems such as those arising in computer networks.

What it takes to move from an adaptive system to a *complex* adaptive system is an open question and one that can engender endless debate. At the most basic level, the field of complex systems challenges the notion that by perfectly understanding the behavior of each component part of a system we will then understand the system as a whole. One *and* one may well make two, but to really understand two we must know both about the nature of "one" and the meaning of "and."

The hope is that we can build a *science of complexity* (an obvious misnomer, given the quest for simplicity that drives the scientific enterprise, though alternative names are equally egregious). Rather than venturing further on the well-trodden but largely untracked morass that attempts to define complex systems, for the moment we will rely on Supreme Court Justice Stewart's words in his concurring decision on a case dealing with obscenity (*Jacobellis v. Ohio*, 1964): "I shall not today attempt further to define the kinds of material I understand to be embraced within that shorthand description; and perhaps I could never succeed in intelligibly doing so. But I know it when I see it."

The field of complex systems must direct its "flight from wonder" toward discoveries that "make the wonderful and complex understandable and simple." We hope that there is a complex systems equivalent of Newton's Laws of Motion that will one day make our current computer simulations appear to us as archaic as machines implementing Ptolemy's epicycles. Even when the fundamental laws of complex adaptive social systems are uncovered, however, it is unlikely that our flight from wonder will be complete. Knowing Newton's Laws of Motion reveals a key simplicity in the world around us, and while we may take delight in the power of so simple an idea to explain the motion of our universe, the thrill of the discovery quickly wanes with the mundaneness of the outcome. Laws emerging from complex adaptive systems have an entirely different character—knowing Darwin's theory of evolution in no way diminishes the wonder that ensues as we observe its implications.

Writings on complexity in the social sciences go back hundreds of years, with Adam Smith's *The Wealth of Nations* (1776) representing one of the earliest and most cohesive discussions of the topic (see figure 1.1). One of the prime drivers of economic theory over the past two centuries has been Smith's concept of an "invisible hand" leading collections of self-interested agents into well-formed structures that are no part of any single agent's intention. Although much theoretical progress has been made on this idea, for example, the elegant proofs of existence given by Arrow and Debreu or the various contributions based on fanciful mechanisms like Walrasian auctioneers, the actual mechanisms behind the invisible hand still remain largely, dare we say, invisible.

Indeed, the tools and ideas that have been developed over the past decade hint at a new world of scientific possibilities for understanding complex adaptive social systems. While our ability to theorize about social systems has always been vast, the set of tools available for pursuing these theories has often constrained our theoretical dreams either implicitly or explicitly. Smith faced few limits while writing about the complexity of the world around him, whereas Arrow and Debreu's existence proof required a much more constrained view of social behavior. Often, tools get mistaken for theories with unfortunate consequences; elaborate computer programs (perhaps with lovely graphics) or mathematical derivations are occasionally assumed to make a real scientific statement, regardless of their scientific underpinnings. Indeed, entire literatures have undergone successive refinements and scientific degradation, during each generation of which the original theoretical

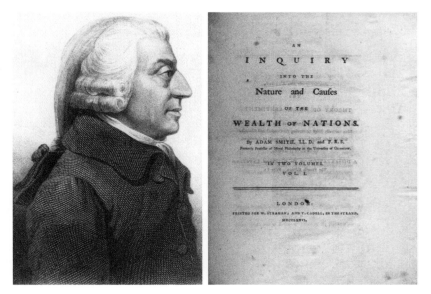

Figure 1.1. Adam Smith and *The Wealth of Nations* (1776). Smith's writings represent one of the earliest coherent descriptions of complexity in social systems.

notions driving the investigation are crowded out by an increasing focus on tool adeptness. This often results in science that is "smart but not wise."

Using traditional tools, social scientists have often been constrained to model systems in odd ways. Thus, existing models focus on fairly static, homogeneous situations composed of either very few or infinitely many agents (each of whom is either extremely inept or remarkably prescient) that must confront a world in which time and space matter little. Of course, such simplicity in science is a virtue, as long as the simplifications are the right ones. Yet, it seems as though the world we wish to know lies somewhere in between these extremes.

One of the most powerful tools arising from complex systems research is a set of computational techniques that allow a much wider range of models to be explored. With these tools, any number of heterogeneous agents can interact in a dynamic environment subject to the limits of time and space. Having the ability to investigate new theoretical worlds obviously does not imply any kind of scientific necessity or validity— these must be earned by carefully considering the ability of the new models to help us understand and predict the questions that we hold most dear.

The science of complex systems is a rapidly evolving area, in terms of both domains and methods. The interest in this area, as well as its rapid subsequent diffusion, has been rather remarkable (especially in a field like economics, where, as Paul Samuelson (1999, xi) once remarked, "science advances funeral by funeral"). We intend for this book both to summarize some key past contributions as well as to lay out an agenda for the future. Any such agenda will require the efforts of many scientists, and we hope to provide sufficient insights and practical guidance so that others can productively join in this research effort.

The tools and ideas emerging from complex systems research complement existing approaches, and they should allow us to build much better theories about the world when they are carefully integrated with existing techniques. Some of the discussions in this book surround basic issues in good scientific modeling. Having a good understanding of these issues is certainly a prerequisite for anyone interested in pursuing work in this area, and unfortunately explicit discussions of modeling are rarely encountered by most scholars.

The book's central theme, "The Interest in Between," has two meanings. The first relates to the level and techniques we use to illustrate the core material in complex adaptive social systems. The second concerns the scientific space that this area occupies.

Complex systems has become both a darling of the popular press and a rapidly advancing scientific field. Unfortunately, this creates a gap between popular accounts that rely on amorphous metaphors and cutting-edge research that requires a technical background. Here we hope to provide a point of entry that lies between metaphor and technicalities. Our work focuses on simple examples that are accessible, yet also contain much deeper foundational insights. This approach is analogous to learning game theory by studying the Prisoner's Dilemma or the Centipede game. While game theory rests on a very abstract and technical foundation—fixed points, hemicontinuous correspondences, and the like—most of the core insights are contained in the analysis of these simple games. In a similar spirit, here we rely on simple models and examples to convey the key ideas. These illustrations will exist in between metaphor and abstract mathematics, in between the flowery language that has taken hold in the press and concrete computations. We view this "in-between" as a good point of entry into the material and hope that it gives readers the ability and interest to dig deeper into the field as they see fit.

We have strived to make this book accessible to both academics and the sophisticated lay reader. Whether you are a graduate student or faculty member in the social sciences trying to understand better what complex systems is about and how it could be used, an engineer hoping to improve your models of processes by using social agents, or someone interested in business, economics, or politics who wants a deeper understanding of the causes and implications of complexity, you should find this book useful and approachable.

Ultimately the study of complex systems illuminates the interest in between the usual scientific boundaries.

It is the interest in between various fields, like biology and economics and physics and computer science. Problems like organization, adaptation, and robustness transcend all of these fields. For example, issues of organization arise when biologists think about how cells form, economists study the origins of firms, physicists look at how atoms align, and computer scientists form networks of machines.

It is the interest in between the usual extremes we use in modeling. We want to study models with a few agents, rather than those with only one or two or infinitely many. We want to understand agents that are neither extremely brilliant nor extremely stupid, but rather live somewhere in the middle.

It is the interest in between stasis and utter chaos. The world tends not to be completely frozen or random, but rather it exists in between these two states. We want to know when and why productive systems emerge and how they can persist.

It is the interest in between control and anarchy. We find robust patterns of organization and activity in systems that have no central control or authority. We have corporations—or, for that matter, human bodies and beehives—that maintain a recognizable form and activity over long periods of time, even though their constituent parts exist on time scales that are orders of magnitude less long lived.

It is the interest in between the continuous and the discrete. The behavior of systems as we transition between the continuous and discrete is often surprising. Many systems do not smoothly move between these two realms, but instead exhibit quite different patterns of behavior, even though from the outside they seem so "close."

It is the interest in between the usual details of the world. We need to find those features of the world where the details do not matter, where large equivalence classes of structure, action, and so on lead to a deep sameness of being.

The science of complex systems and its ability to explore the interest in between is especially relevant for some of the most pressing issues of our modern world. Many of the opportunities and challenges before us— globalization, sustainability, combating terrorism, preventing epidemics, and so on—are complex. Each of these domains consists of a set of diverse actors who dynamically interact with one another awash in a sea of feedbacks. To understand, and ultimately to harness, such complexity will require a sustained and imaginative effort on the part of researchers across the sciences.

Kenneth Boulding summarized science as consisting of "testable and partially tested fantasies about the real world." The science of complex systems is not a new way of doing science but rather one in which new fantasies can be indulged.

CHAPTER 2

Complexity in Social Worlds

> I adore simple pleasures. They are the last refuge of the complex.
>
> —*Oscar Wilde, The Picture of Dorian Gray*

> When a distinguished but elderly scientist states that something is possible, he is almost certainly right. When he states that something is impossible, he is very probably wrong.
>
> —*Arthur C. Clarke, Report on Planet Three*

WE ARE SURROUNDED by complicated social worlds. These worlds are composed of multitudes of incommensurate elements, which often make them hard to navigate and, ultimately, difficult to understand. We would, however, like to make a distinction between complicated worlds and complex ones. In a complicated world, the various elements that make up the system maintain a degree of independence from one another. Thus, removing one such element (which reduces the level of complication) does not fundamentally alter the system's behavior apart from that which directly resulted from the piece that was removed. Complexity arises when the dependencies among the elements become important. In such a system, removing one such element destroys system behavior to an extent that goes well beyond what is embodied by the particular element that is removed.

Complexity is a deep property of a system, whereas complication is not. A complex system dies when an element is removed, but complicated ones continue to live on, albeit slightly compromised. Removing a seat from a car makes it less complicated; removing the timing belt makes it less complex (and useless). Complicated worlds are reducible, whereas complex ones are not.

While complex systems can be fragile, they can also exhibit an unusual degree of robustness to less radical changes in their component parts. The behavior of many complex systems emerges from the activities of lower-level components. Typically, this emergence is the result of a very powerful organizing force that can overcome a variety of changes to the lower-level components. In a garden, if we eliminate an insect the vacated niche will often be filled by another species and the ecosystem will continue to function; in a market, we can introduce new kinds of traders and remove old traders, yet the system typically maintains its ability to set sensible prices. Of course, if we are too extreme in such changes, say, by eliminating a keystone species in the garden or all but one seller in the market, then the system's behavior as we know it collapses.

When a scientist faces a complicated world, traditional tools that rely on reducing the system to its atomic elements allow us to gain insight. Unfortunately, using these same tools to understand complex worlds fails, because it becomes impossible to reduce the system without killing it. The ability to collect and pin to a board all of the insects that live in the garden does little to lend insight into the ecosystem contained therein.

The innate features of many social systems tend to produce complexity. Social agents, whether they are bees or people or robots, find themselves enmeshed in a web of connections with one another and, through a variety of adaptive processes, they must successfully navigate through their world. Social agents interact with one another via connections. These connections can be relatively simple and stable, such as those that bind together a family, or complicated and ever changing, such as those that link traders in a marketplace. Social agents are also capable of change via thoughtful, but not necessarily brilliant, deliberations about the worlds they inhabit. Social agents must continually make choices, either by direct cognition or a reliance on stored (but not immutable) heuristics, about their actions. These themes of connections and change are ever present in all social worlds.

The remarkable thing about social worlds is how quickly such connections and change can lead to complexity. Social agents must predict and react to the actions and predictions of other agents. The various connections inherent in social systems exacerbate these actions as agents become closely coupled to one another. The result of such a system is that agent interactions become highly nonlinear, the system becomes difficult to decompose, and complexity ensues.

2.1 THE STANDING OVATION PROBLEM

To begin our exploration of complex adaptive social systems we consider a very simple social phenomenon: standing ovations (Schelling, 1978; Miller and Page, 2004). Standing ovations, in which waves of audience members stand to acknowledge a particularly moving performance, appear to arise spontaneously.[1] Although in the grand scheme of things

[1]There are circumstances, such as the annual State of the Union address before the U.S. Congress, where such behavior is a bit more orchestrated.

standing ovations may not seem all that important, they do have some important parallels in the real world that we will discuss later. Moreover, they provide a convenient starting point from which to explore some key issues in modeling complex social systems.

Suppose we want to construct a model of a standing ovation. There is no set method or means by which to do so. To model such a phenomenon we could employ a variety of mathematical, computational, or even literary devices. The actual choice of modeling approach depends on our whims, needs, and even social pressure emanating from professional fields.

Regardless of the approach, the quest of any model is to ease thinking while still retaining some ability to illuminate reality.

A typical mathematical model of a standing ovation might take the following tack. Assume an audience of N people, each of whom receives a signal that depends on the actual quality of the performance, q. Let $s_i(q)$ give the signal received by person i. We might further specify the signal process by, say, assuming a functional form such as $s_i(q) = q + \epsilon_i$, where ϵ_i is a normally distributed random variable with a mean of zero and standard deviation of σ. To close the model, we might hypothesize that in response to the signal, each person stands if and only if $s_i(q) > T$, where T is some critical threshold above which people are so moved by the performance that they stand up and applaud.

Given this simple mathematical model, how much of reality can we illuminate? The model could be used to make predictions about how many people would stand. We could tie this prediction to key features of the model; thus, we can link the elements like the quality (q) of the performance, the standing threshold (T), and even the standard deviation of the signal (σ) to the likelihood of an initial ovation of a given size. Given the current form of the model, that is about the extent of what we can predict. These predictions do provide some illumination on reality, but they fail to illuminate some of the key elements that make this problem so interesting in the first place (like the waves of subsequent standing).

Given this, we might want to amend the model to shed a bit more light on the subject at hand. It is probably the case that people respond to the behavior of others in such situations. Therefore, we can add a parameter α that gives the percentage of people who must stand for others to ignore their initial signals and decide to stand up regardless. In some fields, like economics, we might even delve a bit deeper into the notion of α and see if we can tie it to some first principles, for example, perhaps people realize that their signals of the performance are imperfect and thus they update them using the information gathered by observing the behavior of others. We will avoid such complications here and just assume that α exists for whatever reason.

Our elaborated model provides some new insights into the world. If the initial group of people standing exceeds α percent, then everyone will rise; if it falls short of this value, then the standing ovation will remain at its initial level. Again, we can tie the elements of the model to a prediction about the world. By knowing the likelihood of various-sized initial ovations, we can predict (given an α) the likelihood of everyone else joining the ovation.

As clean and elegant as the mathematical model may be, it still leaves us wanting some more illumination. For example, we know that real ovations do not behave in the extreme way predicted by the model; rather, they often exhibit gradual waves of participation and also form noticeable spatial patterns across the auditorium. In the model's current form, too much space exists in between what it illuminates and what we want to know about the real world.

To capture this additional illumination, we might extend the mathematical model even further by using ideas from complex systems. This approach may require us to model using a different substrate, most likely indirect computation rather than direct mathematics, but for the moment this choice is less important than the directions we wish to take the modeling. The first elaboration we could undertake is to place each person in a seat in the auditorium, rather than assuming that they attend the theater on the head of a pin. Furthermore, we might want to assume that people have connections to one another, that is, that people arrive and sit with acquaintances (see figure 2.1).[2]

Once we allow people to sit in a space and locate next to friends, the driving forces of the model begin to change. For example, the initial assumption of independent signals is now suspect. It is likely that people seated in one part of the theater (or "side of the aisle") receive a different set of signals than others. Locations not only determine physical factors, such as which other patrons someone can see, but also may reflect a priori preferences for the performance that is about to begin. Similarly, in an audience composed of friends and strangers, people may differentially weight the signals sent by their friends, either because of peer pressure or because the friendships were initially forged based on common traits.

Assuming that individuals now have locations and friends introduces an important new source of heterogeneity. In the mathematical model, the only heterogeneity came from the different draws of ϵ_i. Now, even "identical" individuals begin to behave in quite different ways, depending on where, and with whom, they are seated.

---

[2] We once had a group of economics graduate students model the standing ovation. Not one of them allowed the possibility of people attending the theater with acquaintances. We hope this is more a reflection of how economists are trained than of how they live.

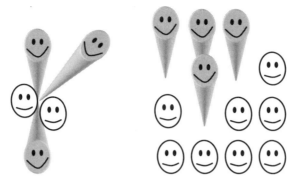

Figure 2.1. Two views of modeling the standing ovation. In its simplest form, the model requires that everyone shares the same seat in the auditorium (*left*), while the more elaborate model (*right*) allows space, friendship connections, and physical factors like vision to play a vital role in the system. While the simple model might rely on traditional tools like formal mathematics and statistics, the more elaborate model may require new techniques like computational models using agent-based objects to be fully realized.

The dynamics of the model also becomes more complicated. In the original model, we had an initial decision to stand, followed by a second decision based on how many people stood initially. After this second decision, the model reached an equilibrium where either the original group remained standing or everyone was up on their feet. The new model embodies a much more elaborate (and likely realistic) dynamics. In general, it will not be the case that the model attains an equilibrium after the first two rounds of updating. Typically, the first round of standing will induce others to stand, and this action will cause others to react; in this way, the system will display cascades of behavior that may not settle down anytime soon.

These two modeling approaches illuminate the world in very different ways. In the first model either fewer than α percent stand or everyone does; in the second it is possible to have any percentage of people left standing. In the first model the outcome is determined after two periods; in the second cascades of behavior wash over the auditorium and often reverberate for many periods. In the first model everyone's influence is equal; in the second influence depends on friendships and even seat location. Oddly, the people in the front have the most visual influence on others yet also have the least visual information, whereas those in the back with the most information have the least influence (think of the former as celebrities and the latter as academics).

The second model provides a number of new analytic possibilities. Do performances that attract more groups lead to more ovations? How does changing the design of the theater by, say, adding balconies, influence ovations? If you want to start an ovation, where should you place your shills? If people are seated based on their preferences for the performance, say, left or right side of the aisle or more expensive seats up front, do you see different patterns of ovations?

Although standing ovations per se are not the most pressing of social problems, they are related to a large class of important behaviors that is tied to social contagion. In these worlds, people get tied to, and are influenced by, other people. Thus, to understand the dynamics of a disease epidemic, we need to know not only how the disease spreads when one person contacts another but also the patterns that determine who contacts whom over time. Such contagion phenomena drive a variety of important social processes, ranging from crime to academic performance to involvement in terrorist organizations.

2.2 What's the Buzz?

Heterogeneity is often a key driving force in social worlds. In the Standing Ovation problem, the heterogeneity that arose from where people sat and with whom they associated resulted in a model rich in behavioral possibilities. If heterogeneity is a key feature of complex systems, then traditional social science tools—with their emphases on average behavior being representative of the whole—may be incomplete or even misleading.

In many social scenarios, differences nicely cancel one another out. For example, consider tracking the behavior of a swarm of bees. If you observe any one bee in the swarm its behavior is pretty erratic, making an exact prediction of that bee's next location nearly impossible; however, keep your eye on the center of the swarm—the average—and you can detect a fairly predictable pattern. In such worlds, assuming behavior embodied by a single representative bee who averages out the flight paths of all of the bees within the swarm both simplifies and improves our ability to predict the future.

2.2.1 Stay Cool

While differences can cancel out, making the average a good predictor of the whole, this is not always the case. In complex systems we often see differences interacting with one another, resulting in behavior that deviates remarkably from the average.

To see why, we can return to our bees. Genetic diversity in bees produces a collective benefit that plays a critical function in maintaining hive temperature (Fischer, 2004). For honey bees to reproduce and grow, they must maintain the temperature of their hive in a fairly narrow range via some unusual behavioral mechanisms. When the hive gets too cold, bees huddle together, buzz their wings, and heat it up. When the hive gets too hot, bees spread out, fan their wings, and cool things down.

Each individual bee's temperature thresholds for huddling and fanning are tied to a genetically linked trait. Thus, genetically similar bees all feel a chill at the same temperature and begin to huddle; similarly, they also overheat at the same temperature and spread out and fan in response.

Hives that lack genetic diversity in this trait experience unusually large fluctuations in internal temperatures. In these hives, when the temperature passes the cold threshold, all the bees become too cold at the same time and huddle together. This causes a rapid rise in temperature and soon the hive overheats, causing all the bees to scatter in an over ambitious attempt to bring down the temperature. Like a house with a primitive thermostat, the hive experiences large fluctuations of temperature as it continually over- and undershoots its ideals.

Hives with genetic diversity produce much more stable internal temperatures. As the temperature drops, only a few bees react and huddle together, slowly bringing up the temperature. If the temperature continues to fall, a few more bees join into the mass to help out. A similar effect happens when the hive begins to overheat. This moderate and escalating response prevents wild swings in temperature. Thus, the genetic diversity of the bees leads to relatively stable temperatures that ultimately improve the health of the hive.

In this example, considering the average behavior of the bees is very misleading. The hive that lacked genetic diversity—essentially a hive of averages—behaves in a very different way than the diverse hive. Here, average behavior leads to wide temperature fluctuations whereas hetero- geneous behavior leads to stability. To understand this phenomenon, we need to view the hive as a complex adaptive system and not as a collection of individual bees whose differences cancel out one another.

2.2.2 Attack of the Killer Bees

We next wish to consider a model of bees attacking a threat to the hive.[3] Some bees go through a maturation stage in which they guard the

---

[3]This is a simplified version of models of human rioting constructed by Grannoveter (1978) and Lohmann (1993). Unlike the previous example, the direct applicability to bees is more speculative on our part.

entrances to the hive for a short period of time. When a threat is sensed, the guard bees initiate a defensive response (from flight, to oriented flight, to stinging) and also release chemical pheromones into the air that serve to recruit other bees into the defense.

To model such behavior, assume that there are one hundred bees numbered 1 through 100. We assume that each bee has a response threshold, R_i, that gives the number of pheromones required to be in the air before bee i joins the fray (and also releases its pheromone). Thus, a bee with $R_i = 5$ will join in once five other bees have done so. Finally, we assume that when a threat to the hive first emerges, R bees initiate the defensive response (to avoid some unnecessary complications, let these bees be separate from the one hundred bees we are watching). Note that defensive behavior is decentralized in a beehive: it is initiated by the sentry activities of the individual guard bees and perpetuated by each of the remaining bees based only on local pheromone sensing.

We consider two cases. In the first case, we have a homogeneous hive with $R_i = 50.5$ for all i. In the second case, we allow for heterogeneity and let $R_i = i$ for all i. Thus, in this latter case each bee has a different response threshold ranging from one to one hundred. Given these two worlds, what will happen?

In the homogeneous case, we know that a full-scale attack occurs if and only if $R > 50$. That is, if more than fifty bees are in the initial wave, then all of the remaining one hundred will join in; otherwise the remaining bees stay put. In the heterogeneous case, a full-scale attack ensues for any $R \geq 1$. This latter result is easy to see, because once at least one bee attacks, then the bee with threshold equal to one will join the fray, and this will trigger the bee with the next highest threshold to join in, and so on.

Again, notice how average behavior is misleading. The average threshold of the heterogeneous hive is identical to that of the homogeneous hive, yet the behaviors of the two hives could not be more different. It is relatively difficult to get the homogeneous hive to react, while the heterogeneous one is on a hair trigger. Without explicitly incorporating the diversity of thresholds, it is difficult to make any kind of accurate prediction of how a given hive will behave.

2.2.3 Averaging Out Average Behavior

Note that the two systems we have explored, regulating temperature and providing defense, have very different behaviors linked to heterogeneity. In the temperature system, heterogeneity leads to stability. That is, increased heterogeneity improves the ability of the system to stabilize on a given temperature. In the defense system, however, heterogeneity induces instability, with the system likely to experience wild fluctuations in response to minute stimuli.

The difference of response between the two systems is due to feedback. In the temperature system, heterogeneity introduces a negative feedback loop into the system: when one bee takes action, it makes the other bees less likely to act. In the defense system, we have a positive feedback loop: when one bee takes action, it makes the other bees more likely to act.

2.3 A TALE OF TWO CITIES

To explore further the modeling of complexity, we consider a simple world composed of two towns, each of which has three citizens. Furthermore, we assume that each town has to make a choice about an important public issue: whether to serve its citizens red or green chile at its annual picnic. Citizens possess preferences over chile and strongly prefer one type over the other.[4] To make the analysis interesting, we assume that two of the citizens in each town prefer green to red chile while the remaining person prefers the opposite.

Though stark, this scenario builds from an extensive literature in the social sciences on the allocation of public goods and services to citizens (Samuelson, 1954; Tiebout, 1956). Public goods and services flow across all members of society without exclusion or diminution once offered. Moreover, as we will see, the model also touches on even deeper issues surrounding the decentralized sorting of agents within a complex adaptive system.

Before we can explore the behavior of the model, we need to define two further elements. The first is how does a town, given a set of citizens, select what chile to offer. The second is how do citizens react to the choices of the towns.

A town could use several mechanisms to decide what type of chile to offer. It could employ a dictator, flip a coin, or implement some other political process, such as majority rule. For the moment, we will assume that each town uses majority rule. Given this scenario, majority rule implies that each town will always offer green chile (two votes to one). Note that this outcome is not ideal, as one citizen in each town always ends up consuming her less-preferred meal (see figure 2.2).

---

[4]For those who enjoy both, New Mexican restaurants offer the option of ordering your chile "Christmas."

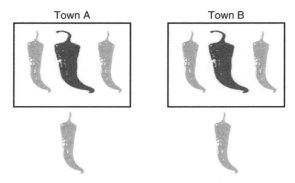

Figure 2.2. A symmetric Tiebout world. Here two towns each have three citizens, two of whom prefer green to red chile. Both towns currently offer green chile at their annual picnic. Given this scenario, the system is at an equilibrium, even though two of the citizens are not getting their favorite chile.

Now, suppose we give our citizens some mobility, that is, any citizen is free to switch towns if she so desires. We assume that citizens will move only if the alternative town is offering a better meal. If each town is serving green chile, no citizen has any incentive to relocate and everyone stays put.

Yet, something should be done. The current situation possesses a tragic symmetry that prevents the red chile lovers from every realizing their favored outcome since they are always the minority in either town. To improve this situation, we must find a way to break the symmetry.

One way to break the symmetry is to introduce some randomness into the system. For example, we could have one citizen randomly decide to move to the other town for whatever reason. If this citizen is a red chile lover, then the town she vacated is left with two green chile lovers and her new town now has two people who like red and two who like green chile. Instead, if the citizen that relocates is a green chile lover, then the vacated town is left with one of each type, while the other town now has three green and one red chile lover. Notice that regardless of who moves, we are always left with one town that is strongly green chile and one that has equal numbers of each type.

Given this situation, we would expect that eventually the town with a split vote will offer red instead of green chile. Once this occurs, we now have one town offering red and one offering green chile. The symmetry is now broken, and the citizens in each town can immediately re-sort themselves and self-select the town that perfectly meets their chile needs. This leaves one town offering green chile populated by four green chile lovers and one town offering red chile with two red chile lovers, and all

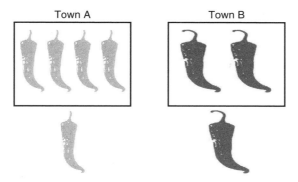

Figure 2.3. Broken symmetry in the Tiebout world. Once the two towns offer different types of chile—perhaps due to noise in the political system—the citizens will immediately re-sort themselves. The system again attains an equilibrium, though in this case each citizen now gets her favorite type of chile. Note that this new equilibrium is much more robust to minor perturbations than the former one.

of the citizens would be worse off if they moved (see figure 2.3). This latter configuration is quite stable to random moves of individuals, as a single citizen moving will not alter the majority in either town.

An alternative way to break the symmetry is to alter slightly the behavioral rules that control our citizens. Suppose that agents are willing to relocate if they can at least maintain their level of happiness (rather than improve it). Such a change in behavior allows for what biologists call *neutral mutations*, that is, movements in the underlying structure that do not directly impact outcomes. Even though neutral mutations do not have an immediate effect, they can lead to better outcomes eventually by changing what is possible. In the initial case, any of the citizens is willing to move since both towns offer the same type of chile. Regardless of who moves, one town is always left with a split vote, and the symmetry breaking we saw previously is again possible.

The system demonstrates some key features of complex adaptive social systems. First, we have a web of connections that, in this case, results from citizens linking to one another by being resident in a given town. Second, we see change induced by choices made by all of the different types of agents in the system. Citizens must decide where to move, and towns must decide what type of chile to offer. Moreover, the system as a whole must "decide" how to sort the citizens among the towns, although this latter "choice" is not a conscious calculation of the system per se, but rather an implicit computation resulting from the decentralized choices made by each citizen and town. The model also demonstrates how a social system can get locked into an inferior outcome and how, with the introduction of noise or different behavioral rules, it can break out of such outcomes and reconfigure itself into a better arrangement.

The model also incorporates other key themes in complex adaptive social systems: equilibria, dynamics, adaptation, and the power of decentralized interactions to organize a system. The system has multiple equilibria, some of which are inferior to others. The key dynamics that occur in the model are the choice dynamics of each town induced by the voting system and the movement dynamics of each citizen implied by her preferences and each town's offerings. Note that these dynamics imply that towns adapt to citizens, while citizens also adapt to towns. Finally, we see how the system's dynamics result in local, decentralized behaviors that ultimately organize the citizens so that their preferences align with other citizens and each town's offerings align with its residents.

2.3.1 Adding Complexity

While our model gives us some useful intuitions and insights, it is also (quite intentionally) very limited. Like all good models, it was designed to be just sufficient to tell a story that could be understood easily yet have enough substance to provide some insights into broader issues. Moving beyond the limitations of this model is going to require some compromises—namely, if we want to expand the potential for insights, we will likely need to complicate the model and, perhaps, muddy the analytic waters.

For example, suppose we wish to explore more fully Tiebout's (1956) concept of "voting with your feet." That is, can we characterize better the ability of social systems to sort citizens dynamically among towns? The simplifications in the preceding model were rather drastic; we had two towns, six citizens, a single issue (choice of chile), and a single mechanism to determine what each town offered (majority rule). If we wish to go beyond any of these constraints, we will quickly start to run into trouble in pursuing the thought experiment framework used previously.

In economics, formal modeling usually proceeds by developing mathematical models derived from first principles. This approach, when well practiced, results in very clean and stark models that yield key insights. Unfortunately, while such a framework imposes a useful discipline on the modeling, it also can be quite limiting. The formal mathematical approach works best for static, homogeneous, equilibrating worlds. Even in our very simple example, we are beginning to violate these desiderata. Thus, if we want to investigate richer, more dynamic worlds, we need to pursue other modeling approaches. The trade-off, of course, is that we must weigh the potential to generate new insights against the cost of having less exacting analytics.

One promising alternative approach is the development of computation-based models. In the Tiebout system, through computation we can allow multiple towns and citizens, as well as more elaborate preference and choice mechanisms. Thus, we can consider a world in which each town must make binary choices over multiple issues, such as whether to, say, serve red or green chile at the annual picnic, allow smoking in public places, and set taxes either high or low. Once we admit multiple issues, our citizens will need to have more complicated preference structures to account for the more elaborate set of choices. This will imply that, instead of just two types of citizens, we now have a much more heterogeneous population. Finally, instead of using majority rule as the sole means by which a town chooses its offerings, we can admit a variety of other possible social choice mechanisms. For example, towns might use a form of democratic referenda where, like simple majority rule, citizens get to vote on each issue and the majority wins; or perhaps the towns could rely on political parties that develop platforms (positions on each possible choice) and then vie for the votes of the populace.

Rather than fully pursuing the detailed version of the model we just outlined (interested readers should see Kollman, Miller, and Page, 1997), here we provide just an overview. Using computation, we can explore a world with multiple issues, citizens, towns, choices, and choice mechanisms. For example, consider a model where each town must make binary decisions across eleven issues. Each citizen has a preference for each issue that takes the form of a (randomly drawn) weight that is summed across all of the choices in a town's platform to determine the citizen's overall happiness. Of particular interest at the moment is the effectiveness of different public choice mechanisms in allocating citizens to towns and towns to platforms.

We will allow towns to use a variety of choice mechanisms to determine what they will offer. At one extreme we can employ *democratic referenda* (essentially majority rule on an issue-by-issue basis), while at the other we will consider a party-based political processes whereby political parties propose platforms and then compete with one another for votes. In this latter mechanism, we can consider worlds with two or more parties, either where the winning party takes all in *direct competition* (that is, the winning party's platform is what the town offers) or where, in a system of *proportional representation*, the final platform offered by the town is a blend, weighted by votes, of each individual party's platform.

Again, we impose a simple dynamic on the system: the citizens in a town, mediated by the choice mechanism, determine what the town will offer across the eleven issues and, once that is determined, citizens look around and move to their favorite town based on their own preferences and each town's current offerings. We iterate this process multiple times and ultimately investigate the final match of citizens to towns and towns to issues. For the moment, we judge each mechanism only by its effectiveness at maximizing the overall happiness of the citizens after a fixed amount of time. Thus, a good outcome will have citizens with similar preferences living in the same town, and that town offering a platform that aligns well with the preferences of its, relatively homogeneous, residents.

To get our bearings, first consider the case of a world with only a single town. In such a world the dynamic implied by citizens moving from town to town is nullified, and the only dynamic element of the model is that arising from the town altering its offerings via the choice mechanism. Thus, the best outcome will depend on the ability of the choice mechanism to come up with a platform that closely matches the preferences of the population. We find that, under these conditions, democratic referenda lead to the best outcome, followed by two political parties competing under direct competition, then multiple parties with proportional representation, and finally more than two parties using direct competition. Under democratic referenda, the system immediately locks into the median position of the voters on each issue; under the other mechanisms, party competition can result in the town's platform changing from period to period and not necessarily achieving the median on any one issue. Under the preference structure of our model, the median voter position on each issue will typically maximize the overall welfare of a fixed group of citizens confined to *a single town*. Therefore, democratic referenda are the best mechanisms for maximizing social welfare in a world consisting of only a single town.

Oddly, when we allow additional towns into the system, democratic referenda no longer lead to the highest social welfare. In fact, the effectiveness of the different choice mechanisms is completely reversed, and democratic referenda become the worst possible institution rather than the best. (See figure 2.4.)

Why does this happen? Fortunately, computational models are quite amenable to exploring such questions; in essence, we have a laboratory on the desktop and can systematically propose, test, and eliminate key hypotheses to understand better the outcomes we are observing.

To develop some needed intuition, consider the following. If we are interested in maximizing the overall happiness of our citizens with multiple towns, we must achieve two ends. First, we need to sort

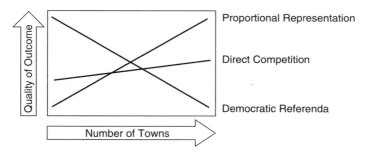

Figure 2.4. Results of a computational Tiebout model. As we increase the number of towns in the system, the effectiveness of the different choice mechanisms in achieving high social welfare completely reverses.

the citizens among the available towns so that citizens with similar preferences reside in the same town. Without such a sorting, the social welfare generated by each town will be compromised given the diversity of wants. Second, each town must choose across the issues so as to maximize the happiness of its residents. As noted, democratic referenda are very effective at deriving a stable platform of choices that maximizes happiness for a given town. Given this observation, the failure of democratic referenda with multiple towns must be related to their inability to sort adequately the citizens among the towns.

A deeper investigation into the dynamics of the system confirms that the mechanisms other than democratic referenda result in far more initial movement of the citizens among the towns. Democratic referenda tend to stabilize the system quickly, freezing the citizens and platforms in place after only a few iterations. That is, after only a few rounds each town is offering a fixed platform, and no citizen wants to move. The other mechanisms are much more dynamic, in the sense that the platforms of each town keep changing during the early periods and the citizens tend to migrate much more often. Eventually, even these latter systems settle down to a state with little platform change and few migrations.

Earlier we saw how noise in the system allows it to break out of inferior sortings and to lock into superior ones. Of course, noise alone is not sufficient to guarantee a quality sorting of the citizens— to achieve high levels of social welfare, you need the noise to result in relatively homogeneous groups of citizens in each town and each town implementing platforms that approach something akin to the median issue positions across the local voters.

In fact, the choice mechanisms that work best in our more complicated model have a subtle, but key, property. These mechanisms tend to introduce noise into the system when the local citizens' preferences are heterogeneous and to reduce this noise as the citizens become more homogeneous. Thus, if the citizens in a given town have very different preferences from one another, the more successful mechanisms will tend to induce more sorting. As the local citizens become more and more similar, these same mechanisms tend to converge on something approaching the median position on each issue. The notion that good political mechanisms should have such an inherent design is somewhat intuitive: if everyone in a district wants the same thing, the mechanism should deliver it; if, on the other hand, there is a diversity of wants, then the political process should jump around among the various options.

This "natural" annealing process turns out to be a very effective way to promote the decentralized sorting of citizens among towns. To achieve the highest social welfare, we need homogeneous collections of citizens in each town receiving roughly the median policy of the local residents. When the overall sorting of the system is poor, that is, when the mix of citizens in each town tends to be heterogeneous rather than homogeneous, then we should introduce a lot of noise into the platforms. Such noise will induce some citizens to migrate, and this migration will often cascade across other towns and result in a fairly large-scale resorting of the citizens. However, as the citizens become better sorted, that is, as each town becomes more homogeneous, the choice mechanisms should "cool" (anneal) the system by stabilizing on platforms that closely match the relatively homogeneous preferences of each town's citizens.

The notion of annealing to improve the structure of decentralized systems was first recognized a few thousand years ago in early metal-working. Heating metal tends to disrupt the alignment of (add noise to) the individual atoms contained in a metal; then, by slowly cooling the metal, the atoms can align better with one another, resulting in a more coherent structure. Kirkpatrick, Gelatt, and Vecchi (1983), based on some ideas from Metropolis et al. (1953), suggested that "simulated" annealing could be used as an effective nonlinear optimization technique. Thus, the Tiebout model shows how different institutions (here, public choice mechanisms) can become natural annealing devices that ultimately result in a decentralized complex adaptive social system seeking out global social optima.

By pursuing the more elaborate computational model, we achieved a number of useful ends. First, we were able to investigate some important new questions, such as the impact of citizen heterogeneity, multiple towns, and differing choice mechanisms on the ability of a system to achieve high social welfare. Second, the more elaborate model provided some new insights into how such systems behave, the most important being the idea that well-structured noise can jolt a system out of inferior equilibria and lead it toward superior ones, and that choice mechanisms can be designed to introduce such noise in a decentralized way. This intuition is contrary to our usual way of thinking about such problems. Noise is usually considered to be a disruptive force in social systems, resulting in perturbations away from desirable equilibria rather than a means by which to attain them.

The complex-systems approach also allows us to explore the system's robustness. The system autonomously responds to all kinds of changes. We can randomly change the preference profiles for some of the citizens, introduce or remove issues, and so on. In each case, the system will adapt to these changes by presenting new platforms and inducing new migrations. Depending on the rate of change, we may see the system slowly moving through a sequence of equilibria or find ourselves with a world constantly in flux.

Although we have focused our discussion on a political system allocating public goods, the basic ideas embodied in the model are much broader. Decentralized sorting arises across a variety of domains. For example, workers seek jobs, traders match with trading partners, individuals form social groups and clubs, and industries sort out standards and geographic locations. All of these scenarios could be cast as decentralized sorting problems similar to the one just discussed. Moreover, we could use the ideas developed here to formulate new kinds of decentralized sorting algorithms that could be used to, say, sort computer users across resources (like servers) or on-line communities (like bulletin boards or tagging).

The Tiebout world we have explored is a nice example of a much broader quest. There is nothing that is unique about the Tiebout world in terms of its complexity. Like most social systems, it displays some dynamics, heterogeneity, and agent interactions that, even in vastly simplified models, can easily introduce complexity. Even a little bit of complexity implies that the conventional tools we often employ to investigate the world will be limited in their ability to yield insights and prescriptions. We are not claiming that these more conventional tools are useless; indeed, they are an important complement in any quest to understand the world.[5] The computational approach pursued here provided a number of new directions and insights that both enhanced, and was enhanced by, more conventional techniques.

---

[5] In the example presented, the investigation of the system first began with the more elaborate computational model. Based on that experience, we were able to develop the thought experiment with which we opened this section.

2.4 New Directions

The notion that real social systems often result in complex worlds is nothing new. More than two hundred years ago Adam Smith described a world where the self-interested social behavior of butchers, brewers, bakers, and the like resulted in the emergence of a well-defined order. While social science has been able to develop tools that can help us decipher some parts of this system, we have yet to understand fully the inner workings of the world around us. Unfortunately, we are at the mercy of a world characterized by change and connections, and thus our ability to make sense of our world is often undermined by the same characteristics that make it so fascinating and important.

The application of computational models to the understanding of complex adaptive social systems opens up new frontiers for exploration. The usual bounds imposed by our typical tools, such as a need to keep the entire model mathematically tractable, are easily surmounted using computational modeling, and we can let our imagination and interests drive our work rather than our traditional tools. Computational models allow us to consider rich environments with greater fidelity than existing techniques permit, ultimately enlarging the set of questions that we can productively explore. They allow us to keep a broad perspective on the multiple, interconnecting factors that are needed to understand social life fully. Finally, they give us a way to grow worlds from the ground up and, in so doing, provide a viable means by which to explore the origins of social worlds.

As we move into new territory, new insights begin to spill forth. Sometimes these insights are strong enough to stand on their own; at other times, they provide enough of a purchase on the problem that we can employ time-tested older techniques to help us verify and illuminate the newly acquired insights. On occasion, of course, computational models leave us with a jumbled mess that may be of no help whatsoever, though, with apologies to Tennyson, 'tis better to have explored and lost than never to have explored at all.

Social science has failed to answer, or simply ignored, some important questions. Sometimes important questions fall through the cracks, either because they are considered to be in the domain of other fields (which may or may not be true) or because they lie on the boundaries between two fields and subsequently get lost in both. More often than not, though, questions are just too hard and therefore either get ignored or (via some convoluted reasoning) are considered unimportant. The difficulty of answering any particular scientific question is often tied to the tools we have at hand. A given set of tools quickly sorts problems into those we could possibly answer and those we perceive as too difficult to ever sort out. As tools change, so does the set of available questions.

Throughout this book, we pursue the exploration of complex systems using a variety of tools. We often emphasize the use of computational models as a primary means for exploring these worlds for a number of reasons. First, such tools are naturally suited to these problems, as they easily embrace systems characterized by dynamics, heterogeneity, and interacting components. Second, these tools are relatively new to the practice of social science, so we take this as an opportunity to help clarify their nature, to avoid misunderstandings, and generally to advance their use. Finally, given various trends in terms of the speed and ease of use of computation and diminishing returns with other tools, we feel that computation will become a predominant means by which to explore the world, and ultimately it will become a hallmark of twenty-first-century science.

2.5 Complex Social Worlds Redux

We see complicated social worlds all around us. That being said, is there something more to this complication? In traditional social science, the usual proposition is that by reducing complicated systems to their constituent parts, and fully understanding each part, we will then be able to understand the world. While it sounds obvious, is this really correct? Is it the case that understanding the parts of the world will give us insight into the whole? If parts are really independent from one another, then even when we aggregate them we should be able to predict and understand such "complicated" systems. As the parts begin to connect with one another and interact more, however, the scientific underpinnings of this approach begin to fail, and we move from the realm of complication to complexity, and reduction no longer gives us insight into construction.

2.5.1 Questioning Complexity

Thus, a very basic question we must consider is how complex, versus complicated, are social worlds. We suspect that the types of connections and interactions inherent in social agents often result in a complex system. Agents in social systems typically interact in highly nonlinear ways. Of course, there are examples, such as when people call one another during the course of a normal day, where agent behavior aggregates in ways that are easily described via simply statistical processes.

Nonetheless, a lot of social behavior, especially with adaptive agents, generates much more complex patterns of interaction. Sometimes this is an inevitable feature of the nature of social agents as they actively seek connections with one another and alter their behavior in ways that imply couplings among previously disparate parts of the system. Other times, this is a consequence of the goal-oriented behavior of social agents. Like bees regulating the temperature of the hive, we turn away from crowded restaurants and highways, smoothing demand. We exploit the profit opportunities arising from patterns generated by a stock market and, in so doing, dissipate their very existence. Like bees defending the hive, we respond to signals in the media and market, creating booms, busts, and fads.

If social worlds are truly complex, then we might need to recast our various attempts at understanding, predicting, and manipulating their behavior. In some cases, this recasting may require a radical revision of the various approaches that we traditionally employ to meet these ends. At the very least, we need to find ways to separate easily complex systems from merely complicated ones. Can simple tests determine a system's complexity? We would like to understand what features of a system move it from simple to complex or vice versa. If we ultimately want to control such systems, we either need to eliminate such forces or embrace them by productively shaping the complexity of a system to achieve our desired ends.

Another important question is how robust are social systems. Take a typical organization, whether it be a local bar or a multinational corporation. More often than not, the essential culture of that organization retains a remarkable amount of consistency over long periods of time, even though the underlying cast of characters is constantly changing and new outside forces are continually introduced. We see a similar effect in the human body: typical cells are replaced on scales of months, yet individuals retain a very consistent and coherent form across decades. Despite a wide variety of both internal and external forces, somehow the decentralized system controlling the trillions of ever changing cells in your body allows you to be easily recognized by someone you have not seen in twenty years. What is it that allows these systems to sustain such productive, aggregate patterns through so much change?

Our modeling of social agents tends toward extremes: we either consider worlds composed of remarkably prescient and skilled agents or worlds populated by morons. Yet, we know that real agents exist somewhere in between these two extremes. How can we best explore this middle ground? A key issue in exploring this new territory is figuring out the commonalities among adaptive agents. While it is easy to specify behavior at the extremes, as we move into the middle ground, we are suddenly surrounded by a vast zoo of curious adaptive creatures. If we are stuck having to study every creature individually, it will be difficult to make much progress, so our underlying hope is that we can find some way to distill this marvelous collection of behaviors down to just a few prototypical ones. Once this is done, we can begin to make progress on a science of adaptive behaviors.

We know that adaptive agents alter the world in which they live. What we do not know is how much agent sophistication is required to do so effectively and what other conditions are necessary for this to happen. In general, the link between agent sophistication and system outcome is poorly understood. Theoretical work in economics suggests that optimizing agents out for their own benefit can, without intention, lead a market system toward efficiency under the right conditions. Moreover, experimental and computational work suggests that such outcomes are possible even with nonoptimizing agents. Ultimately, it would be nice to have a full characterization of the interplay between adaptation and optimality in social systems.

Another realm where we have a limited understanding is the role of heterogeneity in systems. We know that in, say, ecological systems homogeneity can be problematic. For example, using a few genetic lines of corn maximizes short-term output but subjects the entire crop to a high risk of destruction if an appropriate disease vector arises. Homogeneity in social systems may have similar effects. A homogeneous group of agents in, say, a market might result in a well-functioning institution most of the time, but leave the possibility that these behaviors could synchronize in such a way that on occasion the market will crash. By introducing an ecology of heterogeneous traders, such fluctuations might be mitigated. Perhaps heterogeneity is an important means by which to improve the robustness of systems. If so, does this work via complexifying the system or via some other mechanism?

The idea of social niche construction is also important. Agents, by their activities, can often alter the world they inhabit and, by so doing, form new niches. For example, the development of membranes early in the history of life on Earth allowed various biological components to bind together and isolate themselves from the external world. This fundamentally altered their local environment creating new opportunities for interacting with the world. Similarly, the formation of merchant guilds, corporations, and political organizations fundamentally altered both the internal world faced by agents and the external world in which these new entities operated. We would like to know when and how agents construct such niches.

The role of control on social worlds is also of interest. The ability to direct the global behavior of a system via local control is perhaps one of the most impressive, yet mysterious, features of many social systems. In the natural world, tens of thousands of swarm-raiding army ants can form cohesive fronts fifty feet across and six feet deep that can sweep through the forest for prey. This entire operation is controlled via locally deposited chemical signals. At a grander scale, a vast decentralized systems of human markets of all types orchestrate the activities of billions of individuals across the span of continents and centuries. Fully understanding how such decentralized systems can so effectively organize global behavior is an enduring mystery of social science. We do have some hints about how this can happen. For example, adding noise to the system (as we saw in our Tiebout model) may actually enhance the ability of a system to find superior outcomes. We also know that some simple heuristics that arise in some contexts, such as the notion that in a market new offers must better existing ones, result in powerful driving forces that enhance the ability of the system to form useful global patterns.

Every social agent receives information about the world, processes it, and acts. For example, in our Tiebout model, the behavior of the citizens was very straightforward (get information about the offerings of the various towns, process this via your preferences, and act by moving to your favorite town), while that of each town was a bit more elaborate (get information about the preferences of the citizens across the issues, process this via either exact or adaptive mechanisms to develop a new platform, and act by implementing this platform).

Traditional economic modeling tends to have a fairly narrow view of the issues that arise in acquiring information, processing it, and acting. In these models, agents tend to have access to all available information, process it with good fidelity and exacting logic directed toward optimization, and act accordingly. Where traditional economics gains its power is that these restrictions make for relatively easily modeling across a broad spectrum of social activity. Notwithstanding the apparent success of this approach in some domains, one wonders whether such a restricted view of these three elements is appropriate. While clearly these restrictions give us leverage from which to generate insights across a variety of social realms, we also know that in many cases the core tenets driving the approach are misplaced (though it is still an open issue whether this matters in the end). For example, the recent wave of work in behavioral economics is based on the notion that the processing of information by humans may take place in ways that dramatically diverge from the traditional view.

Much of the work we discuss throughout this book relaxes the traditional assumptions about information acquisition, processing, and acting. We want to consider models in which information is selectively acquired across restricted channels of communication. We want to look at agents that process information via adaptive mechanisms or restricted rules rather than exacting logic. We want to explore models in which actions are often limited and localized. How do all of these factors embody social complexity and what does this mean for the practice of social science?

PART II

Preliminaries

The next two chapters are devoted to some more general issues in complex adaptive social systems. The first covers broader issues surrounding scientific modeling. Throughout this book we explore various modeling techniques, and thus having a solid foundation from which to consider such work is a necessity. Unfortunately, such discussions tend to get relegated to the dark arts if they are discussed at all, and we feel that a more explicit treatment of this topic is needed. The second chapter discusses emergence in complex systems. While emergence appears to be a key hallmark of complex systems, explicit discussions are hard to find.

Although the next two chapters provide some nice foundational concepts for readers, the ordering is a bit arbitrary as the full context for these discussions is not developed until later in the book. Thus, some readers may prefer to skip ahead at this point and come back to these chapters at a later time.

CHAPTER 3

Modeling

> For every complex problem, there is a solution that is simple, neat and wrong.
>
> —*H. L. Mencken*

> Things should be made as simple as possible—but no simpler.
>
> —*Albert Einstein*

> Nothing is built on stone; all is built in sand. But we must build as if the sand were stone.
>
> —*Jorge Luis Borges*

WE BEGIN WITH a discussion of the basics of scientific modeling. This topic is so fundamental to the scientific enterprise that it is often assumed to be known by, rather than explicitly taught to, students (with the exception of a high school lecture or two on the "scientific method"). For whatever reasons, learning about modeling is a lot like learning about sex: despite its importance, most people do not want to discuss it, and no matter how much you read about it, it just doesn't seem the same when you actually get around to doing it.

All modeling requires the faith that, as Borges expresses it, we can occasionally turn the sand of the real world into stone. Effective models require a real world that has enough structure so that some of the details can be ignored. This implies the existence of solid and stable building blocks that encapsulate key parts of the real system's behavior. Such building blocks provide enough separation from details to allow modeling to proceed. For example, Mendel was able to develop his key ideas about heredity without knowing anything about DNA, and economists have been able to generate useful theories of individual and firm behavior without having to delve deeply into the human mind or the organization of the firm. This ability to ignore is a crucial component of scientific progress as it allows us, just like the parent trying to stop the endless regress of a three-year-old's "why" questions, to say "just because." Of course, the art of good science is knowing when to say "just because," for if we are able to invoke that incantation at the right moment, the sand underlying our model's foundation will turn to stone.

Systems, whether scientific models or real-world entities, that do not have sufficient underlying structure are very unstable, difficult to understand, and hard to control. Imagine a world in which every detail mattered—even a slight alteration in the world would result in a cascade of changes that resonate throughout all levels of the system. For complex systems (either real or artificial) to be built and maintained, there needs to be some isolation of the component parts, lest "for want of a nail" we lose the kingdom.

The basics of modeling transcend any particular set of tools we may use to create the model. Tools do, of course, direct the modeling as they provide the scaffolding upon which the model must be built. Nonetheless, whether we use the Gedanken experiments of Mach and Einstein or the correspondences of Arrow and Debreu, our judgments about the quality of the model can be decoupled from the particular set of tools used to create it. The remainder of this chapter discusses some fundamentals of modeling, using both an intuitive illustration (road maps) and a more formal derivation (homomorphisms).

3.1 MODELS AS MAPS

One of the best models that we encounter in our daily experience is the road map. Maps allow an enormous range of people to easily acquire, and productively use, information about a complex reality. We can use maps not only for making accurate predictions about how to manipulate the world (for example, to get from point A to B), but also to answer a variety of questions that were not part of the map maker's original intention (for example, to uncover useful patterns of population or geology).

Maps are valuable for a variety of reasons. One reason is that they leave out a lot of *unnecessary* details. In so doing, they minimize distractions and allow us to focus on the questions that we most care about. Good maps are those that have just barely enough details (see figure 3.1). Consider placing all of the possible details of the world (for example, major highways, secondary highways, roadside restaurants, gas stations, cities, elevation contours, waterways, street addresses, economic activity, people, and so on) each on a separate transparency. We could then create custom maps by overlaying various selections from this vast set of details. For a long-distance truck driver, the most useful map might consist of the sheets with major cities, highways, roadside restaurants, and diesel stations. A hiker would want overlays containing elevation contours, waterways, and foot paths.

As we successively add more overlays, the map gains more and more detail at a cost of more distractions and difficulty disentangling useful

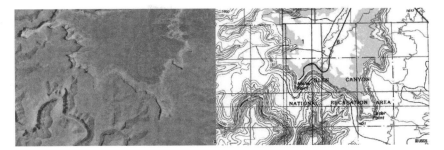

Figure 3.1. Maps as models. A topographic map (*right*) takes the detailed reality of an aerial photograph (*left*) and distills it down to a few key features that match the essential needs of the user. Scientific models must achieve a similar end.

connections. Each time we add an unnecessary overlay, the map loses some of its value. Eventually, the map becomes so complex that it is easier to abandon the stack of overlays and just use the real world itself.[1] Good modeling requires that we have just enough of the "right" transparencies in the map. Of course, the right transparencies depend on the needs of a particular user.

Another desirable feature of a map is that it is easily understandable to others. Maps must communicate their ideas across people and time. To accomplish this task effectively, they not only need to be simple, but users must have a common understanding. Sometimes this understanding is communicated explicitly by, say, the use of legends (which hopefully incorporate some intuitive design features, such as having rest stops being marked by picnic table icons). Often, we also require an implicit understanding on the part of the user. In the case of a map, users are assumed to know that some of the compromises made to create the map require counterintuitive thinking—for example, the shortest route between two points may not be a straight line. Misunderstandings about the implicit knowledge embodied in a map often lead to serious mistakes.

Good maps not only allow us to predict key features of the world, but they also enable us to discover new phenomena. If we look at a map of the world (or, better yet, a globe), theories of continental drift

---

[1] "The Cartographers Guilds struck a Map of the Empire whose size was that of the Empire, and which coincided point for point with it. The following Generations, who were not so fond of the Study of Cartography as their Forebears had been, saw that that vast Map was Useless, and not without some Pitilessness was it, that they delivered it up to the Inclemencies of Sun and Winters." Jorge Luis Borges and Adolfo Bioy Casares, *On Exactitude in Science* (1946), English translation from Jorge L. Borges, *A Universal History of Infamy* (London: Penguin Books, 1975).

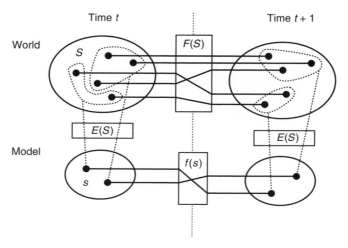

Figure 3.2. A formal model of models. Dotted lines indicate the equivalence class mappings.

seem obvious.[2] Plotting earthquake locations can be used to find hidden fault systems. Maps of navigable water systems or resource locations may suggest likely population centers. Snow's 1855 map of cholera cases in the Soho district of London revealed both the mode of transmission and source (a pump on Broad Street) of the disease. None of these phenomena were anticipated by the original map makers who were solely focused on accurately representing the real world in a compact form; yet all of them (and many more) naturally arise out of the chosen representation.

3.2 A More Formal Approach to Modeling

Here we present a more formal approach to the ideas underlying modeling—a model of modeling. Our discussion relies on the mental modeling ideas developed by Holland et al. (1986) for creating artificial learning systems. We focus on trying to model a world that varies over discrete time steps (though one could obviously apply these ideas to other types of systems). The basic outline of the underlying ideas are presented in figure 3.2. In the top half of the figure, we represent the real world that we are interested in modeling; in the lower half, we depict our model.

We assume that the real world consists of various states, S, and a transition function $F(S)$ that maps a given state at time t into a new state

---

[2]Though it wasn't until around 1800 that Humboldt initiated this line of thinking, and it still took another hundred years before the full hypothesis was introduced.

at time $t + 1$. The function $F(S)$ is unknown to the modeler, and while the modeler would like to uncover $F(S)$, this will typically be impossible given the dimensionality of the state space and potential complexity of $F(S)$. For example, consider the issue of forecasting the weather. The actual state space of this system, S, is enormous consisting of the position and characteristics of all of the relevant atoms. The laws of physics, $F(S)$, tell us how the states of these atoms change from one moment to the next.

As an alternative, the modeler can reduce the size of the state space and seek a simpler transition function. To reduce the size of the state space, the modeler generates equivalence classes: maps from a subset of the real-world states, S, to a model state, s. Here we designate the equivalence class map as $E(S)$. Based on these new model states, the quest of the scientist is to find a useful transition function, $f(s)$, for the model. Continuing our weather example, we might consider using an equivalence class, $E(S)$, that maps configurations of atoms into measures of barometric pressure and humidity. We would then need to find a function, $f(s)$, that can predict how patterns of pressure and humidity are transformed over time. For example, it may be the case that when low- and high-pressure areas abut, the humidity increases and the two pressures begin to equilibrate.

The success of a particular model is tied to its ability to capture the behavior of the real world. Suppose we begin with real-world state S'. The model transforms this state into $s' = E(S')$ and then predicts that we will find ourselves in state $f(s')$ in the next time period. In the real world, state S' becomes state $F(S')$ in the next time period. Thus, the model "coincides" with the real world if $f(E(S')) = E(F(S'))$, that is, if we end up at the same model state regardless of whether we (1) first transform the initial real-world state into its equivalence class and then run it through the model's transition function, or (2) first allow the real-world state to be transitioned to its next state and then map this state, via the equivalence class, to the model. The requirement that the maps between the model and the real world must be commutative in this way is known as a *homomorphism*. Thus, the goal of modeling under this view is to find a set of equivalence classes and a transition function that results in a useful homomorphism.

Consider the earlier problem of modeling the weather. Rather than worrying about the position and characteristics of every atom in the atmosphere (an impossible task to be sure), we first use equivalence classes to collapse this space down to, say, measures of pressure and humidity. We next develop notions of how patterns of pressure and humidity are transformed over time. This model has a homomorphism if on any given day we can predict the pressure and humidity in the next time period knowing the pressure and humidity now *and* that prediction matches the actual pressure and humidity observed.

A model requires choices of both the equivalence classes and the transition function, and the art of modeling lies in judicious choices of both. For any given real-world problem, there are likely to be multiple equivalence mappings (and associated transition functions) that will result in homomorphisms. The value of any particular set of choices depends on the current needs of the modeler. Moreover, the difficulty of discovering the model's transition function will be closely tied to the chosen equivalence mapping, and thus modelers must make trade-offs between these two elements. Choosing an overly broad set of equivalence classes simplifies the task of finding an appropriate transition function, $f(s)$, leading to a homomorphism, but at the cost of lowering the model's resolution and value. In the limit, consider the useless but homomorphic model in which all possible states of the real world are mapped into a single equivalence class and the model's transition function is the identity.

Finding useful homomorphisms is often difficult in practice, and modelers may be willing to forgo perfect homomorphism. Even models in which a few states of the world are "exceptional" may still have a lot of value. Oftentimes the exceptions have little consequence for the main applications of the model. Furthermore, if the number of exceptions is small, the old model can be preserved by refining its equivalence classes and transition function so that every exception is handled individually. If, however, the number of exceptions becomes too large, the model must undergo a more radical change. Such changes range from simple modifications of the underlying features of the model (with, perhaps, a corresponding loss of resolution) to a much more dramatic reframing entailing an entirely new set of equivalence classes and transition function.

This continual chasing of the "ideal" model results in a Schumpeterian cycle of scientific creative destruction. Modelers attempt to reduce the world to a fundamental set of elements (equivalence classes) and laws (transition functions), and on this basis they hope to better understand and predict key aspects of the world. The ever present quest for refining old, and discovering new, ways to represent the world drives the process of scientific creative destruction.

3.3 MODELING COMPLEX SYSTEMS

The proceding principles of modeling are key to developing a scientific understanding of a system, whether it is simple or complex. One view of modeling complex systems, which (at least implicitly) is held by many

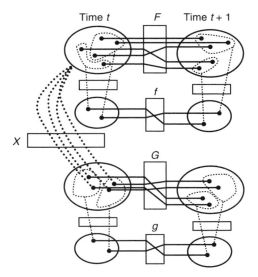

Figure 3.3. Modeling complex systems. The behavior of the entities at one level in the world (*upper panel*) might result in new entities emerging (via function X) that take on new types of behaviors that require a new class of models (*lower panel*).

scientists, is the reductionist hypothesis. This hypothesis suggests that, if we can just get the right simplifications in the model, we will understand everything—if true, then the world around us, including the social world, is "just particle physics." That is, once we understand the basics of particle physics, we can simply apply this knowledge to unravel whatever higher-level systems we may wish to reveal (moving, roughly in order, through physics, chemistry, biology, psychology, sociology, economics, and so on), and knowledge is simply "applied" physics. Dennett (1995) refers to this idea as "greedy reductionism." Of course, as pointed out by Anderson (1972), the fallacy here is that the reductionist hypothesis does not imply a "constructionist" one. Even if we know the fundamentals of a particular system, we may not be able to use that knowledge to reconstruct higher-level systems. It may be, as Anderson says, that "the whole becomes not only more than but very different from the sum of its parts."

The potential implications of this "more is different" hypothesis for modeling complex systems are shown in figure 3.3. The upper part of the panel is identical to that of figure 3.2, and it shows a real system and its associated model. The lower part of the panel represents a higher-level system. Each element of this new system is derived from interactions among the entities in the low-level system. For example, at some point the weather patterns of barometric pressure and humidity discussed previously might result in a completely new entity being formed—like a hurricane—that behaves in a very different way. In the social sciences, the two systems involved might be psychology and economics. Here, individual choice behavior in the low-level psychological system could aggregate into a very different *homo economicus* or firm entity in the higher-level economic system (via the unknown transformation function, X). The new entities that are formed by the lower-level ones interact in a very different world, governed by a new transition function, G. As in the lower-level system, one can begin to model this new system as shown in the bottom part of the figure. However, note that knowledge of the old system (say, knowing F or the details of the old model) does not directly help us model the new, higher-level system. This latter view of the world coincides with the idea of "hierarchical reductionism" put forth by Dawkins (1976).

If more is different, then there is plenty of "fundamental" work at all levels of the scientific enterprise. Each time we move to a new level, we are confronted with a new world that requires new models. Moreover, creating a theory about how these new levels arise from existing ones, namely understanding the function X, becomes important. We would like to be able to develop a theory that helps us understand how states of the world (composed of lower-level entities and interaction rules) are transformed into higher-level entities. Some initial work on this topic has been done with cellular automaton models, where it has been shown that under some conditions a variety of seemingly different interaction rules imply only a few distinct types of high-level behavior (Wolfram, 2002).

3.4 Modeling Modeling

Throughout this book we explore a variety of systems using methodologies ranging from computation to mathematics to thought experiments. Regardless of the system or methodology, our goal is to employ high-quality models. Thus, we apply the same standards of simplicity and elegance to our computational models that we do to our mathematical ones. Models need to be judged by what they eliminate as much as by what they include—like stone carving, the art is in removing what you do not need. Even though a computational model may require thousands of lines of code, if done well it can still embody the simplicity and elegance that is demonstrated in a mathematical model existing in only a few equations.

Having an explicit awareness of the issues surrounding quality modeling is important if we want to work on the frontiers of science. This awareness disciplines our efforts as we explore new problems and employ novel techniques. Creating a model is much like trying to solve a brain teaser. Finding such solutions is often an extremely difficult task involving a combination of theory, practice, and a bit of art. Yet, once discovered, the answer has strong intuitive appeal and appears all too obvious.

CHAPTER 4

On Emergence

He intends only his own gain, and he is in this, as in many
other cases, led by an invisible hand to promote an end which
was no part of his intention.

—Adam Smith, Wealth of Nations

Any sufficiently advanced technology is indistinguishable
from magic.

—Arthur C. Clarke, Profiles of the Future

MUCH OF THE FOCUS of complex systems is on how systems of interacting agents can lead to emergent phenomena. Unfortunately, emergence is one of those complex systems ideas that exists in a well-trodden, but relatively untracked, bog of discussion. The usual notion put forth underlying emergence is that individual, localized behavior aggregates into global behavior that is, in some sense, disconnected from its origins. Such a disconnection implies that, within limits, the details of the local behavior do not matter to the aggregate outcome. Clearly such notions are important when considering the decentralized systems that are key to the study of complex systems. Here we discuss emergence from both an intuitive and a theoretical perspective.

The notion of emergence has deep intuitive appeal. Consider for the moment standing up close to a pixelated picture (see figure 4.1).[1] While each individual pixel can be easily understood in terms of its shape, color, hue, and other properties, it is typically impossible to figure out the entire image by simply scanning across the pixels at close range. As the observer moves back, there is some point at which the overall image begins to resolve, and the pixels become indistinguishable. Once the image has resolved, we can typically make many possible alterations to individual pixels and still not impact the overall image. Indeed, depending on the image, certain types of global pixel properties, such as color, may not even be be needed to have a good sense of the final image.

We may see emergence at many levels. For example, instead of having each pixel composed of a single solid color, we could replace it with a tiny picture whose overall properties can approximate the key

---

[1] For the more romantic among you, assume a stained glass window.

Figure 4.1. Emergence from a mosaic. While the properties of each tile are easy to understand at close range, the true nature of the full image is impossible to comprehend from such information. It is only when you view the mosaic from far away that emergence allows the entire image to become viable.

visual attributes of the previous pixel.[2] Of course, each of these tiny photographs, each emerging from its own set of pixels, could stand by itself. Thus, there is the possibility of multiple layers of emergence, where pictures become pixels that become pictures that become pixels and so on. This may lead us to a "Horton-Hears-A-Who" theory of the universe, in which the world is composed of stacked layers of emergence.

The notion of emergence at many levels is an important one, as each level of emergence can serve as a convenient point at which to dissect the larger system and attempt to understand some of its secrets. Indeed, the boundaries of modern science rely on this property—for example, physics resolves into chemistry, which resolves into biology, which resolves into psychology, which resolves into economics, and so on. Each new science is able to exploit the emergence that is attained by the previous level.

While this metaphor of emergence is very appealing, it leaves open the question of how it should fit into scientific discourse. Part of the innate appeal of emergence is the surprise it engenders on the part of the observer. Many of our most profound experiences of emergence come from those systems in which the local behavior seems so entirely disconnected from the resulting aggregate as to have arisen by magic, echoing Clarke's observation about advanced technology. Examples of such dramatic disconnects include photomosaic pictures, the order and persistence of beehives and foraging ant colonies through simple sets of localized signals, and the stability of a market price generated by the often chaotic and heterogeneous efforts of traders.

[2]The technique of photomosaic pictures exploits this idea.

Alas, surprise and ignorance are closely related. It could be that emergent behavior is simply reflective of scientific ignorance rather than some deeper underlying phenomenon. What may start out as a mystical emergent phenomenon, such as planetary motion prior to Kepler, may turn out to be something rather simple—in the case of Kepler, *just* an ellipse. If all such scientific conundrums can be easily resolved, then perhaps it is true that all of our world is *just* physics. Nonetheless, whether our fascination with emergent phenomena is driven by ignorance or a more profound scientific mystery matters little. Profound scientific mysteries often get resolved in such a way that our prior ignorance becomes apparent, yet it is the ignorance that drives the quest for understanding.

4.1 A Theory of Emergence

To move forward on the scientific exploration of emergence, it is useful to consider what types of theoretical ideas are possible in this area. As we have discussed, emergence is a phenomenon whereby well-formulated aggregate behavior arises from localized, individual behavior. Moreover, such aggregate patterns should be immune to reasonable variations in the individual behavior. Ideally, what we would like to develop are theorems about such phenomena, and, fortunately, at least one such theorem has existed since the early 1700s.

The theorem, the Law of Large Numbers (and its various offshoots, including the Central Limit Theorem), was developed by statisticians over the past few hundred years. It is of interest because it provides some relatively general conditions under which a certain type of aggregate behavior can emerge from the stochastic, microlevel actions of individual agents. Suppose that each individual agent's behavior is summarized by a random variable, X. Furthermore, assume that these variables are mutually independent, have a common distribution, and a mean equal to μ. According to the Law of Large Numbers, the probability that the mean will differ from μ by less than some arbitrary amount tends to one as we increase the number of agents in the system.

Thus, in such systems there is a stable, aggregate property (here the expected value of the common distribution) that emerges from aggregating the activities of the agents. Moreover, this aggregate property is robust to many underlying assumptions about the agents. In the foregoing case of the Law of Large Numbers, the only restriction is that the common distribution has mean μ; other than that, we can vary any of its other characteristics and still maintain the identical aggregate behavior.

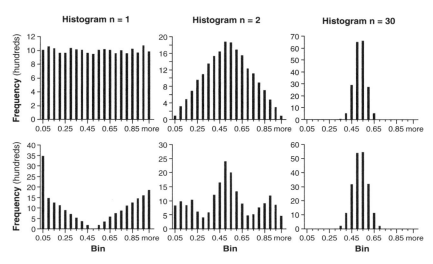

Figure 4.2. Central Limit Theorem. Notwithstanding the form of the initial distribution underlying the random process, the distribution of the mean of the variables generated by this process converges to a Normal distribution as we take larger and larger samples. In the top panel we start with a Uniform distribution, while in the bottom one we begin with a V-shaped distribution. In either case, a Normal distribution "emerges" as the sample size increases. Thus, the macrobehavior resulting from the aggregation of a remarkably diverse set of potential microbehaviors results in a very robust and predictable outcome—a hallmark of emergence.

The Central Limit Theorem provides another example of such a result (see figure 4.2). If we add the assumption that the variance of the common distribution is finite, then the distribution of the average (our aggregate property) will converge to a well-known "normal" form. The remarkable implication of this theorem is that, for an amazing variety of underlying agent behaviors, the global behavior that emerges is described by a simply specified, common form.

As Coates (1956) points out, without these laws, much of the behavior of the social worlds we live in would fall apart. Various activities, ranging from driving on the highway to enjoying the outdoors, would either be excessively crowded or notably desolate at the strangest times, stores and restaurants would run out of the oddest things, life insurance companies and telephone systems would fail, and so on.

The emergence theorems provide useful descriptions of a certain type of complexity that Weaver (1958), in a deeply prescient article, called *disorganized complexity*. The Law of Large Numbers works because as we add more and more independent agents to the world, the vagaries of the stochastic elements, quite literally, average out. With only a few agents, these stochastic elements make it impossible to predict with any certainty the aggregate behavior because individual variation overwhelms any potential predictability, but as we increase the number of agents involved in the world, individual variations begin to cancel one another out, and systemwide predictions become possible.

The key feature of disorganized complexity is that the interactions of the local entities tend to smooth each other out. In the case of the Law of Large Numbers, an unusually high value for one random value is compensated for by an unusually low value of another. Thus, while it is difficult to predict the point at which, say, a particular rain drop meandering down a roof will fall into the gutter, it is easy to predict the activity at any point of the gutter during a rainstorm, as the various meanderings of the drops tend to disrupt one another sufficiently as they flow down the roof so as to spread out the water in a predictable way. Similarly, while predicting the motion of a planet surrounded by a few neighbors is difficult, it is easy to calculate its motion when it is among a sea of other planets, as the various gravitational forces that come into play begin to cancel one another out, and soon only the mean force becomes important. Other phenomena, ranging from population genetics to physical properties like temperature and pressure, also fall within the realm of disorganized complexity.

Thus, in cases of disorganized complexity, it should be easy to derive fairly precise emergence theorems based on fundamental concepts that are centuries old. Unfortunately, disorganized complexity accounts for only one part of our world.

4.2 Beyond Disorganized Complexity

Consider a picture of a face that is composed of black and white pixels. In such a picture, the pixels have relationships to one another that are quite important if we are to recognize the face. Some changes to the picture will not cause us to "lose" the face; for example, having a few of the pixels randomly change color, or even allowing some neighboring pixels to switch places, will preserve the "face." Even some radical changes may not impact our ability to perceive the face, such as altering the color of related pixels (think Warhol) or just showing the important edges of the photograph (see figure 4.3). Moreover, if we are careful, we may be able to "capture" the image with just a few carefully drawn lines, as is done in caricature drawings.

Nonetheless, while we can make some slight changes to the pixels (or even some carefully designed radical changes) and still maintain the

Figure 4.3. Beyond disorganized complexity. The essence of the photograph remains robust to a variety of radical changes. All of these transformations keep intact the relationship of key parts of the original photograph (*left*).

image, doing anything more is likely to destroy the image. As we start to impact more and more pixels (by either randomly altering their colors or allowing neighbors to switch places), we quickly descend into the realm of disorganized complexity. In such a world, the photograph quickly resembles the white noise we see on the television (at least prior to the advent of late-night infomercials) when stations end their broadcast day. While it would be possible to construct the usual disorganized-complexity emergence theorems about, say, the average tone of the picture or the likelihood of an eyelike shape arising somewhere in the photograph, such theorems would fail to capture the essence of the problem of understanding how a decentralized set of pixels can emerge into a familiar face.

Thus, disorganized complexity, while often useful, leaves out a lot of interesting complexity-related phenomena. Disorganized-complexity emergence theorems can be used to calculate the vanishingly small probability that a room full of apes randomly banging away at typewriters will come up with *Hamlet*. Of course, a close relative of an ape did write *Hamlet*, but obviously not by randomly placing words to parchment and hoping for the best. Similarly, while disorganized-complexity theorems can be used to predict the life-span of a human body or a beehive composed of individual agents (cells or bees, respectively), they do not provide any insight into how the various communication and behavioral pathways among the individual agents are able to aggregate into these larger-scale organizations that survive and have behaviors on scales that are completely different than their constituent parts.

Explorations of complex systems have begun to identify the emergent properties of interacting agents—for want of a better term, *organized complexity*. We often see unanticipated statistical regularities emerging in complex systems. These regularities go beyond the usual bounds covered by Central Limit Theorems and such. In chapter 9 we explore a model of sand piles in which we randomly drop grains of sand onto a table. A pile forms as the sand falls, and eventually grains begin to run off the edges of the table in avalanches of various sizes. The distribution of avalanche sizes follows a power law that implies behavior that is quite different from that arising from a normal distribution.

Agent intention can also alter the patterns that emerge in complex systems. In the case of the Sand Pile model, if we give the falling grains of sand a bit of control on where they land and some desires (like maximizing the size of the resulting avalanche), the system is no longer governed by a power law and instead enters a bizarre periodic cycle. As we give agents even more strategic ability, we often see elaborate dances of strategies, with good and bad epochs, cycles, and crashes.

In systems characterized by the Central Limit Theorem, interactions cancel one another out and result in a smooth bell curve. In complex systems, interactions reinforce one another and result in behavior that is very different from the norm.[3] The complex phenomena that arise in physical systems (like earthquakes, floods, and fires) and social ones (like stock market crashes, riots, and traffic jams) are decidedly not "normal," nor are the patterns that emerge as we see birds flock, fish school, and pedestrians follow sidewalks demarcated by invisible traffic lanes.

4.2.1 Feedback and Organized Complexity

When interactions are not independent, feedback can enter the system. Feedback fundamentally alters the dynamics of a system. In a system with negative feedback, changes get quickly absorbed and the system gains stability. With positive feedback, changes get amplified leading to instability.

For example, consider a world in which we have one hundred consumers, each of whom must choose to shop at one of two identical grocery stores. In a world ruled by the Central Limit Theorem, a consumer would choose a store with probability one-half. Thus, each store could expect to see fifty customers on average, though the actual number that showed up would be subject to random variation. In fact, given the underlying process just described, we know that a given store will have, say, more than sixty customers only about 2 percent of the time.

Now, allow customers to act more purposefully and interact with one another. Suppose that customers prefer to be in less-crowded stores. Such

---

[3] One only needs to look at the failure of Long-Term Capital Management in 1998 to realize the practical importance of this distinction. The world in which Long-Term Capital Management played was one governed by fat-tailed distributions, not the Central Limit Theorem.

an assumption introduces a feedback into the system, whereby customers who find themselves in the crowded store begin to shop at the other store. To avoid some odd system-level behavior, we allow only a single customer per period to make such a decision.[4] Given this assumption, in very short order the number of shoppers in each store equilibrates at fifty. Even if we include small external shocks to the system, for example, two customers from different stores take a liking to one another and begin to shop together, the system as a whole will quickly resettle back to the stable configuration with exactly fifty people in each store. Thus, the desire to avoid crowding by each individual induces a negative feedback on the system, resulting in a very stable and predictable outcome.

Agent interactions can also introduce positive feedback into the system. Suppose the same group of one hundred people, also does some banking each day. Imagine that each person has some chance, say 50 percent, of going to the bank and withdrawing money. The bank has limited reserves to cover withdrawals, and thus if too many people withdraw their money, the bank will be unable to cover the claims and become insolvent, causing depositors to panic and demand their money. If the bank has 60 percent reserves, then, as we saw earlier, around 2 percent of the time the bank will go insolvent, resulting in an unfortunate "large event" and an all-out run on the bank.

In our three worlds we see very different behavior. In the first, customers act independently and ignore one another, so the resulting number of customers shopping at a given store is nicely approximated by a normal distribution with a mean of fifty and standard deviation of five. In the second, where customers avoid crowding, we get a degenerate distribution with each store having exactly fifty customers each day. Finally, with the potential for panic, the number of customers arriving at the bank looks identical to the normal distribution we saw in the first case when we have less than sixty customers, but once we hit sixty, all of the remaining weight of the distribution shifts to the right, and we get a fat upper tail.

The contrasts between these images are startling. The world would be a lot easier to understand if we could restrict our attention to the first two scenarios, that is, if agents either avoid direct interaction with one another or interact in such a way that strong negative feedback results in a stable equilibrium. Alas, the vast majority of social science theory focuses on exactly these two types of outcomes. Nonetheless, there are many canonical examples of "large events" that arise in social

---

[4] Without this, there are some dynamics where we can get large swings in the number of customers as they overreact to crowding.

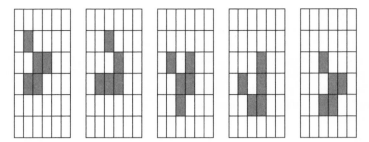

Figure 4.4. Gliders in the Game of Life. A glider in the Game of Life is a configuration of live cells that "moves" across the space. During each successive time step (*left to right*), the set of live cells is altered based on the simple, local rules (see text) of the game. After four time steps, the original configuration of live cells is regenerated, only displaced down and to the right by one cell. If left undisturbed, this structure will continue to "glide" across the space. A more elaborate configuration of live cells, known as a glider gun, is capable of generating such gliders.

systems, such as stock market crashes, riots, outbreaks of war and peace, political movements, and traffic jams. These events are driven by positive feedback, arising from perhaps externalities driven by the behavior of others that change each individual's costs or benefits from acting (for example, as rioting breaks out, your chance of going to jail decreases, and the social benefit of joining in increases) or physical constraints on behavior (such as when the car in front of you on the highway slows down, forcing you to slow down as well to avoid a crash).

Thinking about positive and negative feedback provides only a crude window into the set of possibilities that can emerge in a complex social system. Many complex systems contain both types of feedback. For example, consider Conway's Game of Life. In this game, the world moves in lockstep and is arrayed on a two-dimensional grid, each cell of which can either be dead or alive. A dead cell with exactly three live neighbors is "born" and becomes a live cell next period; otherwise, it remains dead. A live cell with two or three live neighbors "survives" into the next period; otherwise, it dies (either out of "loneliness" or "overcrowding"). Thus, in this system an intermediate amount of life begets life (a positive feedback), while too much or too little life leads to death (a negative feedback). Ultimately, this results in a remarkable set of global patterns in both space and time that can emerge from this simple set of microlevel rules. These patterns are so coherent at times that we can ignore the underlying microlevel rules that generated them and instead rely on the resulting global structures to predict systemwide behavior (see, for example, figure 4.4).

As discussed previously, we have access to some useful "emergence" theorems for systems that display disorganized complexity. However, to fully understand emergence, we need to go beyond these disorganized systems with their interrelated, helter-skelter agents and begin to develop theories for those systems that entail organized complexity. Under organized complexity, the relationships among the agents are such that through various feedbacks and structural contingencies, agent variations no longer cancel one another out but, rather, become reinforcing. In such a world, we leave the realm of the Law of Large Numbers and instead embark down paths unknown. While we have ample evidence, both empirical and experimental, that under organized complexity, systems can exhibit aggregate properties that are not directly tied to agent details, a sound theoretical foothold from which to leverage this observation is only now being constructed.

PART III

Computational Modeling

In the next two chapters we discuss the use of computational models as a theoretical tool, in particular in the modeling of social systems. Over the past decade or so, what was once considered the fringe has become the frontier, and computational models have become much more widely accepted among social scientists. Nonetheless, we feel that it is still useful to outline some of the foundations of this approach, as accessible discussions are hard to find in the literature.

CHAPTER 5

Computation as Theory

By the addition of such artificial Instruments and methods, there may be, in some manner, a reparation made for the mischiefs, and imperfection, mankind has drawn upon it self, by negligence, and intemperance, and a wilful and superstitious deserting the Prescripts and Rules of Nature, whereby every man, both from a deriv'd corruption, innate and born with him, and from his breeding and converse with men, Is very subject to slip into all sorts of errors.

—*Robert Hooke, Micrographia*

The use of computers seems thus not merely convenient, but absolutely essential for such experiments which involve following the games or contests through a very great number of moves or stages. I believe that the experience gained as a result of following the behavior of such processes will have a fundamental influence on whatever may ultimately generalize or perhaps even replace in mathematics our present exclusive immersion in the formal axiomatic method.

—*Stanislaw Ulam, Adventures of a Mathematician*

FOR MANY CENTURIES, houses were constructed by their occupants with perhaps the assistance of a few skilled neighbors. This vernacular architecture led to the creation of unique homes, each reflecting the whims of its builders. Various additions and deletions would accrue over time as the needs of the family changed. Houses were designed with both local materials and conditions in mind. The soundness of such structures was dependent on both luck and the innate engineering skill of each owner—on many occasions a house would collapse due to poor design.

With time, ideas about appropriate home design and construction became less the province of the occupant and more of a professional activity. Ideas like balloon framing (developed in Chicago in the 1830s) allowed the development of sound housing through the use of standardized building materials like two-by-fours. While this new system of building lessened the likelihood of building collapses, it also began to constrain the design choices of architects. The various components had to fit together in particular ways, and alterations required massive reworking of the fundamental elements, not to mention the scorn of the building inspectors. Inevitably, housing design became less concerned with meeting the needs of occupants and local conditions and more focused on realizing structures as cheaply as possible.

The current situation in parts of the social sciences is not unlike that of our intrepid home builders. A long history of both interesting and disastrous vernacular has given way to standardized designs requiring the approval of strict "building inspectors." By using these standard construction techniques, we have been able to build a substantial number of safe structures. But the compromises have been great: we often demand that our occupants fit the structures rather than our structures fitting the occupants, and we celebrate slight alterations of existing plans as architectural prowess. Although the best builders are able to push the standardized components to their structural and aesthetic limits, most of the structures we are building are destined for the most mundane suburbia.

New computational tools offer the possibility of breaking away from the past and forming new theoretical dreams. In architecture, such tools have allowed structures like Gehry's Guggenheim Museum Bilbao to make fanciful dreams a reality. We are not advocating here a return to the vernacular way of building—whatever charm such vernacular has is offset by the likelihood of raising a faulty structure or creating eccentric structures that have little utility beyond their immediate occupants. Rather, we are suggesting that it is time to take our old components and use them in new ways. Such an approach is not without risks, for surely some of the new structures that we build will fall; but others will stand and inspire.

An underlying thesis of this book is that computational modeling is a productive theoretical enterprise. Such an idea often incites controversy. When theoretical computation was introduced into fields like physics, mathematics, and biology, it encountered initial resistance. With time this resistance waned and computational methods became an accepted and integral part of each of these fields. This assimilation process is rapidly ongoing in fields such as economics, where editors who once asked why would one ever want to use computation now suggest shortening papers by removing such discussions since "the computational approach is business as usual."

Notwithstanding the growing acceptance of computation as a theoretical tool, it is still worthwhile to lay out some basic arguments for why we should embrace this approach. As theoretical tools become more widely accepted, their foundations tend to be rarely revisited and even get forgotten by new generations of users. This is unfortunate, as a clear knowledge of the foundations often enhance the resulting theory.

Much of the discussion in this chapter compares computational models to more traditional mathematical modeling techniques, in particular, the neoclassical approach in economics. Our intention here is not to offer a full critique of the traditional approach but instead to ground our discussion on territory that may be familiar to many of our readers. To minimize any misunderstandings at the outset, our view is that tools like mathematics and computation are complements rather than substitutes in the development of sound theory.

5.1 THEORY VERSUS TOOLS

Part of the controversy surrounding computational methods is the result of a confusion between theory and the tools we use to develop theory. A theory is a cohesive set of testable propositions about a phenomenon, and it can be created by employing a variety of tools. The set of tools varies dramatically across, and often within, any given field. For example, economists have relied on theoretical tools that include detailed verbal descriptions, such as Smith's (1776) invisible hand; mathematical analysis, like Arrow's (1951) possibility theorem; and thought experiments, including Hotelling's (1929) railroad line.

Currently, the predominant tool in economic theorizing is the development of mathematical models derived from a set of first principles. The axioms used in such derivations often include assumptions about both the abilities and motives of the underlying agents. Linking these axioms with the notion that social systems tend toward equilibrium states, we can use these models to make predictions. Although there are many reasons to favor such a tool for theory creation, there is no a priori reason to think that this tool should be superior to alternative tools in all (or even most) situations. In fact, economists are often willing to relax the approach when necessary. For example, the theory of supply and demand, probably one of most useful applications of economic thinking, does not currently have a coherent, first-principle basis.

The neoclassical approach has provided a unified and tractable modeling framework from which to attack a variety of interesting social phenomena. These tools provide both a ready set of simplifications for understanding the world and a process by which the implications of these simplifications can be derived in a consistent way. Thus, we can propose a novel problem to a group of neoclassical theorists, send them all off separately to work on the problem, and find, when they reconvene, that their various solutions share a remarkable similarity. Of course, consensus does not necessarily imply correctness. Nonetheless, the possession of a tool that simplifies the development and refinement of theories concerning complex questions is an important step forward in developing a science of society, and it remains one of the most distinguishing hallmarks of economics versus many of its sister social sciences.

Part of the motivation for relying on the neoclassical approach and first-principle derivations is the perception that the testing of social phenomena is extremely difficult, and therefore the logical foundations of the model must be fully secure. Thus, whereas physicists can incorporate unknown constants into their theories because they can eventually determine the values of such constants through careful experimentation, economists can only rely on econometric estimates derived from a few happenstance natural experiments to test their theories. It is argued that social scientists need to be especially sure that the logical foundations of their theories are correct, so that errors, which cannot be easily caught experimentally, will not be propagated.

Whether one agrees with the previous argument, it seems that the important point to be made is not about the value of axiomatic tools but rather the necessity of being able to test our theories. Indeed, recently the area of experimental economics has experienced tremendous growth and acceptance. The legitimacy of using experimental results to challenge existing theories is also gaining acceptance, and it is even starting to drive the theoretical enterprise down new avenues of exploration, such as learning models in games and behavioral economics. While the acceptance of an experimental component to theory creation and testing is likely to continue, it is also the case that experimental methods on human subjects are inherently limited. Some of the systems of most interest to economists, like complex macroeconomic systems composed of hordes of heterogeneous agents, are not easily captured in a standard laboratory setting.[1]

Theories can, and should, be separated from the tools used to derive them. Thus, arguments about whether neoclassical methods are good or bad seem somewhat misplaced. Different tools are good for different things. Some tools, like mathematics, are good for developing precise theories based on simple sets of assumptions. Other tools, like prose, offer the opportunity to explore subtle features of institutions and behavior. Tools need to be judged by their ability to enhance the scientific enterprise; theories need to be judged by how well they are able to improve our understanding of the world around us, and not by what tools we used to derive them.

---

[1] Later, we will argue that computational methods can begin to fill a role similar to that of *E. coli* in biology or *Drosophila* in genetics, serving as a marvelous experimental playground for social theorists.

We often have attachments to particular tools even when they produce the same outcomes. For example, even though the physical appearance of an intricately shaped table leg may be identical regardless of whether it was created by a skilled craftsman using hand gouges or an unskilled worker using a computerized lathe, we may still feel that the leg created by the former method is somehow superior. Such attachments seem misplaced, however, if what we really care about are nice-looking tables. Thus, whether the proposition that countries on a map can always be distinguished through the use of only four colors (the so-called Four-Color Map problem) is proved by the exhaustive enumeration of all possibilities through the use of a computer program (which has been done) or through an elegant (or even nonelegant) axiomatic proof (which has not been done) matters little if all you care about is the basic proposition.

Perhaps we could argue that an axiomatic proof is still superior because it may provide some additional insight into the underlying processes or new theoretical directions in other domains. Whatever the merits of these types of justifications, they implicitly assume that no such insights or directions will be forthcoming from the enumerative approach—an assumption that does not hold in practice.

Alternatively, the argument is made that axiomatic proofs guarantee their outcomes, whereas computational experiments provide only inductive proof.[2] Usually such deductive certainty comes at the cost of being willing to narrow sufficiently the problem domain, so the issue here is under what conditions are we guaranteeing the outcome. If the conditions of the guarantee are excessively onerous, we may well be willing to accept some inexactness in our predictions in return for more favorable circumstances.

A nice illustration of the difference between tools and theories arose many years ago at the Santa Fe Institute. Two researchers, an economist and physicist, were interested in the expected time of discovery in a simplified model of random, bit-wise search (a problem with potential applications in computer science, economics, and genetics). The hour was getting late, so they decided to work on it overnight. The physicist went home and, after spending some time trying to solve the problem analytically, decided to simulate the process on the computer. In short order, he concluded that the answer approximated $n\log n$ plus a constant (where n was the number of bits). The economist spent time deriving an exact solution using recursive function theory. The next day, the economist with the exact solution was quite surprised to hear of

---

[2]Of course, in the case of fully enumerative computations, like the Four-Color Map problem, this argument fails.

the compact approximation derived by the physicist. Both tools were able to "solve" the problem, albeit in different ways, and eventually, using insights realized by both of these techniques, an even simpler way to derive the solution was uncovered.

All tools are designed to simplify some task. If the task we face corresponds with this simplification, then the tool will be of value. On the other hand, if it does not, then, regardless of the quality of the tool, we will be frustrated and the outcome will suffer (even with the best lathe, we will not get good cabriole table legs). Given that most problems are multifaceted, a corollary of this observation suggests that we may need to employ different tools when developing our theories. Thus, a full understanding of supply and demand may require thought experiments using Walrasian auctioneers, axiomatic derivations of optimal bidding behavior, computational models of adaptive agents, and experiments with human subjects.

By attacking problems on numerous fronts, breaches in nature's walls inevitably appear. Though it may be difficult to predict on which front the walls will first crumble, openings on one front are likely to lead to progress on another. Demanding that all attacks take place in a rigorous and prescribed manner is reminiscent of red-coated armies maintaining their formations while confronting the "disorganized" militias of a rebellious colony.

5.1.1 Physics Envy: A Pseudo-Freudian Analysis

During the late nineteenth century, various "cargo cult" societies emerged in the South Pacific. By the mid-twentieth century, inspired by their experiences during World War II, these societies built elaborate mock facilities, such as airstrips and control towers, in hopes of attracting deliveries of goods similar to those that colonial officials once received. Like these societies, we suspect that much of the current view and apparatus of theory in economics is based on misinterpreted observations and misplaced hopes.

There is a commonly held perception in economics that its approach to theorizing closely follows the "one" that is used in physics. Indeed, at certain levels, modern economic theory does resemble some parts of physics, where a small set of well-formulated mathematical models is applied to a broad spectrum of the world. However, based on our interactions at the Santa Fe Institute with a fine group of theoretical physicists, we find that this narrow view of theoretical work is far too restrictive to capture either the reality or the potential of what other fields like physics have to offer in terms of ways to approach theoretical questions in the social sciences.

Theoretical physicists are concerned with, and rewarded by, finding insights about nature through the creation of models and the generation of hypotheses. The emphasis here is on understanding nature, not on the tools used to gain this understanding. Thus, for example, there is a tool used in theoretical physics called the replica method that requires taking the limit of the size of a matrix as it goes to zero. Although this operation has no sensible mathematical justification, the method is popular because of its success in explaining a seemingly disparate set of phenomena. Another theoretical approach uses renormalization groups to reduce complicated stochastic calculations through a recursive series of gross approximations.

It is not that economics is bereft of such tricks, but rather there seems to be a prevailing attitude that such tricks are illegitimate and that unless we have a fully derived-from-first-principles-exact-result, we have failed. Indeed, many of the same theorists that justify "unrealistic" optimization assumptions by invoking Friedman's (1953) arguments that predictions are all that matters seem to ignore this same advice when confronted with new theoretical tools.

The premium in theoretical physics is on gaining insight into interesting phenomena. If the insight is there, then there is little desire for mathematical rigor. Consequently, in physics there is a sharp distinction between the mathematical and theoretical branches. Having a good insight and stating a theorem that is not rigorously proved is acceptable behavior. Once, during a talk at the Santa Fe Institute, a well-known theoretical physicist was asked if he could *rigorously* prove a proposition that he had just made, and his answer was "No, and *I* don't need to, but I'm sure someone can." On first hearing by most economists, this seemingly casual approach to scientific theory is scandalous at best; yet, ultimately it becomes a very productive way to make scientific progress.

While axiomatic rigor is not required for theoretical work in physics, there is still a high premium on good theory—just not on the tools used to develop the theory. Theory must result in insight and withstand testing. Thus, if a theorist decides that some new force governing the interaction between two bodies is likely to be "a lot like gravity," she may model it by claiming that it is approximately like $1/d^r$, where d is distance and r is an unknown parameter. As long as the resulting equation holds up to experimentation, this is a perfectly acceptable theory notwithstanding its lack of a direct, first-principle justification.

There is a branch of physics concerned with mathematical rigor that appears to be fairly separate from the theoretical branch. Sometimes this mathematical branch supplies important clarifications and new theoretical directions, though more often than not its main focus is on taking previous theoretical statements and putting them in a more rigorous context. Perhaps not too surprisingly, the theoretical branch seems to display a fair amount of indifference to this activity, viewing it more as just cleaning up the details. This relationship between mathematics and theory provides an interesting contrast to the norms that have developed in other fields such as economics.

Thus, in physics you can have a theorem that is widely accepted but not rigorously proved. This notion is not exactly alien to economists. For example, the idea of an economic general equilibrium was widely accepted for almost two hundred years before its existence was "proved" by Arrow and Debreu. Notwithstanding this example, economists appear much less willing than their physicist counterparts to accept theorems without complete, first-principle proofs. The ability to theorize without the requirement of axiomatic rigor allows a certain freedom in the attempt to understand nature's mysteries, and when it is exercised well, it can lead to significant advances.

5.2 COMPUTATION AND THEORY

By now, it should be clear that the incorporation of a variety of tools can make for better theory. Much of this book is devoted to the use of computational tools for theory development. Like all tools, computational models have advantages and disadvantages. In the next chapter, we focus on the advantages of computational tools for building theories of complex adaptive social systems. Here, we discuss some general concerns surrounding the use of computational tools.

Like all new tools that are brought into the scientific process, computational models confront a variety of universal concerns: Can these tools generate new and useful insights? How robust are they? What biases do they introduce into our theories? These concerns are obviously important for any type of tool we use in modeling (both traditional and new). Nonetheless, new tools rightfully undergo extraordinary scrutiny in this regard. Eventually, such careful examinations get forgotten or ignored as the tools become an accepted part of scientific practice. For any tool we employ, it is always important that we remember the relevant issues surrounding its appropriate use, regardless of its current level of acceptance.

5.2.1 Computation in Theory

There are many applications of computation *in* theory (versus computation *as* theory), and to avoid confusion we will attempt to distinguish clearly among them (though some fuzziness will remain). First, it is useful to note that the use of a computer is neither a necessary nor a sufficient condition for us to consider a model as computational. Thus, we would not classify using a computer to approximate the integral of an analytic equation as a computational model, nor would we exclude from such a classification Schelling's (1978) coin-based method for analyzing neighborhood segregation.

As discussed in chapter 3, the goal of theory is to make the world understandable by finding the right set of simplifications. Modeling proceeds by deciding what simplifications to impose on the underlying entities and then, based on those abstractions, uncovering their implications. The types of computational models we wish to focus on here are those in which the abstractions maintain a close association with the real-world agents of interest, and where uncovering the implications of these abstractions requires a sequential set of computations involving these abstractions.

The property of close association is often known as "agent-based" modeling, though of course this name is confusing given that most modeling has as its basis the underlying agents. A marginally better term might be modeling using "agent-based objects" versus "abstraction-based objects," with the understanding that agent-based objects require abstractions as well.

Thus, in a neoclassical model of an economic system, the initial "agent" equations are often based on the assumption that individuals optimize their behavior given current information and options. Usually, given mathematical constraints, most of the underlying agents in the real system are subsumed into a single object (a "representative" agent). Finally, the modeling proceeds by manipulating the resulting set of abstract objects and incorporating some additional assumptions about driving forces (for example, the system will seek an equilibrium). Computation is often used in these types of models as a numerical method to help solve a set of abstraction-based objects that defy traditional solution techniques. For example, we may derive a small set of equations that describe the general equilibrium properties of an economy, but even simple sets of such equations may require numerical methods to be solved.

Modeling using agent-based objects proceeds by abstracting the behavior of the individual agents in the system into simplified agents (similar to the optimization step, but less constrained by the need to assume behavior that will anticipate the solution limits of the mathematics to come). Next, collections of these agent-based objects will be "solved" by allowing the objects to interact directly with one another using computation. When computation is applied to such problems, it very much becomes part of the theory.

The second property suggested for our computational models—namely, the need to allow the objects to interact directly with one another— is an interesting requirement. Obviously, there is no a priori reason why models with agent-based objects must be solved using such computations. Such limits may reflect more on the ignorance of the modeler than the needs of the model. Indeed, some of the best-recognized theoretical achievements surround the discovery of simple methods that circumvent difficult computations. Of course, while computational shortcuts are always desirable, the lack of such shortcuts, even if it is due to ignorance on the part of the modeler, should not be viewed as an impediment or diminish the value of computational modeling. Moreover, there are well-known examples of agent-based object models where it can be shown that the shortest way to find the implications of the assumptions is by the full computation itself (Wolfram, 1984a).

The agent-based object approach can be considered "bottom-up" in the sense that the behavior that we observe in the model is generated from the bottom of the system by the direct interactions of the entities that form the basis of the model. This contrasts with the "top-down" approach to modeling where we impose high-level rules on the system—for example, that the system will equilibrate or that all firms profit maximize—and then trace the implications of such conditions. Thus, in top-down modeling we abstract broadly over the entire behavior of the system, whereas in bottom-up modeling we focus our abstractions over the lower-level individual entities that make up the system.

Part of the desirability of the agent-based object approach to modeling surrounds the potential failure of reductionism. As discussed in chapter 3, Anderson (1972) suggested that our traditional view of reductionism may be very misleading when trying to understand complex systems. Suppose we know all of the underlying components of a system and all of the rules by which these components interact—does it then follow that we understand the system? Perhaps not. For example, if we know the color, shape, and location of every piece of glass in a stained glass window, do we necessarily know what figure will emerge from their conglomeration? While clearly all the information is there, we may not be able to imagine what the completed window looks like just from reading about each piece and its location. We may need to "assemble" the window in some form before any kind of image can emerge. We might get a good idea of the window by using a crude line drawing, or perhaps a more elaborate full-color diagram will be required. The level of necessary detail will be linked to some inherent characteristics of the actual image. In some cases, the easiest way to "know" the window is to assemble the entire thing piece by piece.

TABLE 5.1
Computation as Theory

	Simple Structure	*Complicated Structure*
Agent-based objects	Bottom-up modeling (e.g., artificial adaptive agents)	Bottom-up simulation (e.g., artificial life)
Abstraction-based objects	Top-down modeling (e.g., computable general equilibrium)	Top-down simulation (e.g., global warming)

Since the time of Adam Smith we have had a clear understanding of the components of an economic system (self-interested bakers, brewers, and the like) and even a set of interactive rules by which they are governed (market rules and so on). Yet, knowledge of these components does not necessarily imply that we know how prices emerge. Anderson's hypothesis suggests that even if we can fully uncover the microfoundations of behavior—for example, acquire a complete specification of the psychological aspects of behavior or the probability of interaction—we may still not have a simple way to understand their macrolevel implications.

5.2.2 Computation as Theory

Table 5.1 roughly classifies the various applications of computation as theory that are currently in use. The columns attempt to differentiate modeling from simulation. As previously discussed, modeling requires a focus on simple entities and interactions. In addition, good models tend to have a number of other properties: for example, their implications tend to be robust to large classes of changes in the underlying structure, they tend to produce "surprising" results that motivate new predictions, they can be easily communicated to others, and they are fertile grounds for new applications and contexts. As the structure of a model becomes more complicated, many of these desirable features are lost, and we move away from modeling toward simulation. Simulations do have uses in science. They often provide a constructive existence proof of some proposition. Nonetheless, there is a difference between simulation and computational modeling, though as shown in figure 5.1 the exact transition point may be a matter of degree. In the rows of table 5.1, we differentiate between agent- versus abstraction-based objects following the discussion in the previous section.

In the world of computation as theory, we find a variety of ongoing efforts. A main focus of the artificial-life community is the creation of relatively complicated systems of agent-based object models

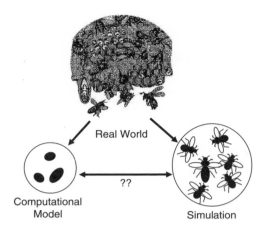

Figure 5.1. Modeling and simulation.

(which we would classify as bottom-up simulations). These are often used as existence experiments to see if a set of rules can imply "lifelike" behavior. Another use of computation is in analyzing large sets of equations for such tasks as projecting world population dynamics or global weather patterns. Each of the equations in these models captures some aspect of system behavior, for example, population dynamics or industrial growth. These top-down simulations use the computer as a way to understand the implications of this set of abstraction-based objects. There are also abstraction-based object systems that rely on much smaller sets of equations, such as much of the work in computable general equilibrium models. These abstraction-based object models use computation to solve the resulting equation system because closed-form mathematical solutions are unavailable or difficult to derive. Finally, there is work that uses agent-based objects (such as artificial adaptive agents) in an attempt to model systems from the bottom up. It is this latter work that is the focus of much of this book.

5.3 Objections to Computation as Theory

While the notion that one can do productive theoretical work using computation models is becoming more widely accepted, it still seems to provoke intense objections by some researchers. Here we attempt to address some of the usual objections to computation as theory. In the next chapter, we take a more proactive view of these methods.

5.3.1 Computations Build in Their Results

A common objection to computation is that the answers are "built-in" to the model, and thus we can never learn anything new from these techniques. Clearly, the first part of this objection is absolutely correct—the results of the computation are built-in since the computer will, without error, follow its predetermined program.[3] Nonetheless, the inference that somehow this makes computation an unacceptable theoretical tool is wrong.

All tools build in answers. If you tell a group of neoclassical theorists to model problem X, we can expect that the answers they derive will be built-in in the sense that each theorist is likely to go off and come up with very similar answers. The real issue is whether the resulting theories are such that the answers are similar due to some undesirable tendency of the tool to embed useless relics or its ability to uncover some deeper truth about nature. Poorly done models, whether computational or mathematical, can always fail because their results are driven by some hidden or obscure black-box feature. Having a premium on clarity and scientific honesty about what is driving the results will always be needed regardless of the theoretical medium.

It is often wrongly assumed that the built-in nature of computational models somehow severely constrains the potential insights that such methods can generate. While, of course, a model can never go beyond the bounds of its initial framework, this does not imply that it cannot go beyond the bounds of our initial understanding (and in so doing allow us to develop new theoretical insights). Hence, even a full knowledge and understanding of Darwin's theory of evolution cannot adequately prepare us for the multitude of wondrous creatures that have resulted from this theory.

To create models that go beyond our initial understanding, we need to incorporate frameworks for emergence. That is, we need to have the underlying elements of the model flexible enough so that new, unanticipated features naturally arise within the model. Some of the most interesting frameworks for emergence are those that create general and flexible structures that get filled in during the course of the computation. For example, we can give a computation access to a general language for programming a variety of strategies to, say, play a game like the Prisoner's Dilemma, and then allow the computation to determine which particular strategy out of this broad class to employ based on some adaptive mechanisms. Such frameworks are rich in possibilities.

---

[3]Note that even when we incorporate stochastic elements, we still must rely on deterministic random-number generators.

By way of an analogy, consider giving students some clay and telling them either to make a coffee mug or an object suitable for drinking a liquid. In both cases, the underlying material (clay) is very flexible, but depending on the instructions we issue we may get very different objects. The instructions to "make a coffee mug" are likely to lead to a set of very similar objects, whereas the request to "make an object suitable for drinking a liquid" could result in a host of possibilities. Useful models arise when we impose just enough instructions to get objects of interest, but not so many as to preimpose a solution.

5.3.2 Computations Lack Discipline

Another common objection to computational models is that they lack sufficient discipline or rigor to be of use. There are a few aspects to this critique. Given the vast potential of computer programs to express a variety of subtle conditions, there are very few constraints on the formulation of models. Mathematical models get around this issue by severely constraining the original formulation of the model, since practitioners know that breaking away from a limited set of assumptions results in an unsolvable model. The lack of constraints on formulating computational models is potentially, of course, a source of great advantage as long as it is used wisely.

Computational approaches, like many other methods, require the modeler to have a high degree of self-discipline to ensure that the techniques are appropriately used. There is, obviously, nothing preventing such a discipline from forming. In many ways, the discipline required for using computational models is similar to that needed in laboratory-based experiments: Is the experiment elegant? Are there confounds? Can it be easily reproduced? Is it robust to differences in experimental techniques? Do the reported results hold up to additional scrutiny?

The hope, of course, is that the inherent nature of computational techniques is such that, even with self-imposed constraints (dictated by quality modeling considerations), new classes of models better able to explain key phenomena will emerge. The flexibility and creativity embodied in computer models often seduce practitioners to continually add features to their work—a practice that must be moderated if good-quality models are to be maintained.

The inherent flexibility of computational models can also make them hard to understand and verify. Mathematical models surmount this issue by having a rigorous set of solution techniques and verification mechanisms. Given the newness of many computational approaches, there has yet to emerge an agreed-upon set of standards. There are a number of techniques that have been, and are being, developed to make sure that computational claims are valid, and we discuss some of these "standards" in appendix B.

The fact that computational models are convenient and flexible should be viewed as a distinct advantage. There are the usual economic trade-offs in using theoretical tools, and where we are given opportunities to acquire "cheap" results, we should be willing to substitute into such tools. Computational models often offer an initial foray into a problem that might be impossible to crack using traditional techniques, though, using the insights from computation, the problem may then yield to more traditional methods.

5.3.3 Computational Models Are Only Approximations to Specific Circumstances

Many analytic methods provide exact answers that are guaranteed to be true. Alas, all models are approximations at some level, so the fact that, say, a mathematical model gives us an exact answer to a set of previously specified approximations may not be all that important. Good answers only make sense when we are asking good questions.

Computational models often result in answers that may be approximations that cannot be directly verified as being correct. Relying on such approximations may be perfectly acceptable, given the potential high cost of getting exact solutions, and even necessary in those cases where exact solutions are infeasible. Moreover, there are techniques, both in how models are formulated and how they are tested, that should help ensure that the results we are finding are not due to some computational anomaly.

Another potential reason for preferring more traditional modeling methods is that they are more generalizable. At the most basic level, a parametric mathematical solution can be used to solve a variety of cases via simple calculations (or other analytic techniques, like comparative statics, can be used to make statements about the influence of the parameters). Bottom-up computational models do not have this feature directly and often must be recalculated each time a new solution is desired. Although this process can be automated, nonetheless it is costly.

Generalizability has a broader interpretation in which a given model can be used to explore new contexts. The ability to use a model for this purpose is tied to the way the model is created, as opposed to the medium of creation. Thus, there are both mathematical and computational models that cannot be easily extended beyond their initial structure and original purpose. Well-designed models capture just enough of the problem to be useful and avoid relying too heavily on any specific implementation details of the theoretical tools used in their creation. Such models are able to flow naturally into other domains, regardless of the modeling substrate.

The ability of a model to generalize is linked closely to its inherent frameworks for emergence. Consider the previous analogy of giving students some instructions for how to form clay. If the instructions are too specific, the types of vessels emerging from the potters will be very constrained—all are likely to have similar form and utility. Under the more general set of instructions, however, the model has the potential to generate a diversity of forms, ranging from sipping bowls to beer steins. Using such results we can distill out the important elements of these worlds (in this case, basins for storing liquid, surfaces for conveying the liquid to the mouth, and so on). Such diversity not only allows us to use the model to explain a broader class of the world, but it also helps us discover the important generic features from which we can begin to build more inclusive theories.

5.3.4 Computational Models Are Brittle

Computational models are often thought to be *brittle*, in the sense that slight changes in one area can dramatically alter their results. This fear is perhaps due to the experience of having a computer program crash after some seemingly innocuous input or alteration. Indeed such crashes are rather dramatic, though they are not unique to computational tools. For example, we see similar collapses in a mathematical model when we alter our assumptions about, say, the form of the utility function or the compactness of the strategy space. For good modeling we need to keep in mind the brittleness of our tools and actively work to avoid producing theories that are too closely tied to any particular assumption.

Brittleness in computational models can be prevented by having a simple and obvious design. The incorporation of emergence frameworks tends to prevent brittleness since such frameworks are typically robust to specific implementation details. Modelers can also prevent brittleness by creating sets of alternative implementations for key features of the model. For example, models that focus on adaptive agents can implement different types of adaptive algorithms. There are also techniques, such as Active Nonlinear Testing (ANTs) (Miller, 1998), where automated searches attempt to uncover brittle areas of the model (see figure 5.2). Some practitioners also use multiple implementations of their models, relying on relatively general specifications and a variety of computer languages to help ensure that specific implementation details do not drive the result. Another approach to prevent brittleness is to "dock" two different models on a common problem, by altering both models until their results converge (Axtell et al., 1996).

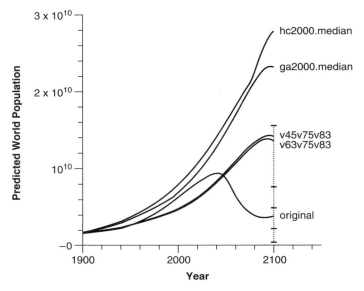

Figure 5.2. Active Nonlinear Testing (ANTs) of a complicated simulation model. An ANTs algorithm was used on a widely publicized simulation model of global population dynamics. The algorithm was allowed to alter any of ninety-six parameters by at most 10 percent each. The algorithm found manipulations that resulted in a predicted population of 28 billion in year 2100 (versus only 4 billion in the original model). This result is well outside the range predicted by a Monte Carlo analysis (given by the dotted line and tick marks at 2100). A more restricted search found that nonlinear interactions among only three parameters could lead to populations of around 14 billion with only minor parametric changes. More details can be found in Miller (1998).

As more and more computational models are developed and explored, our understanding of the typical areas where brittleness may occur will be improved. Such an understanding will not only help us create better models, but it should also be of use in furthering our understanding and control of real systems.

5.3.5 Computational Models Are Hard to Test

Although complex systems and equilibrium models can differ in how they produce testable hypotheses, overall what they produce is quite similar. In both cases we can make refutable predictions.

Consider the familiar model of supply and demand from economics. Equilibrium theory suggests that the market price and quantity will equilibrate where the quantity demanded just meets the quantity supplied. Using this notion, we can do comparative statics. When, say, the price of a substitute commodity rises, we would predict that the demand for the good under consideration would rise at all prices. As a result of this shift in the demand curve, the equilibrium price and quantity should rise.

These comparative static results are powerful. Suppose that we constructed a model of the market for some commodity like apples, and that we considered the effects on equilibrium apple price and quantity of, say, an increase in the production costs of a substitute like pears. Our prediction would be that, if the price of producing a pear rises, the market price of apples and quantity of apples sold would both rise. If we had lots of data on which to base our model, we might even make a point prediction, say, that a 5 percent rise in the price of pears leads to a 3 percent increase in the price of apples and a 2 percent increase in apple sales.

This all seems wonderful and powerful. However, if we looked at actual data to test our model, we would see that things are a bit murkier. The best that we might hope for would be a small cloud in price-quantity space centered on our point estimate. If so, we could consider the difference between what actually happened and what we predicted as noise—as random shocks that we did not include in our model. If, on the other hand, the cloud of data lies far from our point prediction, then we should reject our model.

While the preceding example is a somewhat extreme oversimplification, it does neatly conceptualize empirical tests of equilibrium models. Essentially, we end up comparing points to clouds.

Complex systems models sometimes settle down into equilibrium as well. However, that equilibrium is often not unique, as it may depend on various random elements of the model or nonlinearities. Complex systems models can also remain alive and not settle down to any obvious equilibrium. In these worlds, agents continually respond to the actions of others, and the system is in perpetual motion.

The lack of equilibria in complex systems models does not imply a lack of regularities. Imagine a crude complex systems model of apple and pear markets. In this market, buyers would use rules to decide whether to buy apples or pears depending upon the price. Sellers would use rules to set prices based upon past sales and on the costs they pay to produce apples and pears. Moreover, buyers and sellers might be positioned in space. Buyers would go to particular sellers and only experiment with new sellers with some probability. Neither buyers nor sellers would have elaborate models of prices. Instead, sellers would adjust prices and buyers would adjust purchases by some crude means.

Any given outcome of this model would differ from others depending upon the random features inherent in the rules. This does not mean that the model makes no predictions. If we ran the model a hundred or a thousand times, we would get a distribution of outcomes—a cloud. In this particular instance, that cloud would probably show that an increase in the production costs of pears leads to an increase in the number of apples sold as well as an increase in the price of apples. When we compare the outcome of this model to a real-world situation, we would be comparing the distribution of outcomes produced by the model to another distribution of outcomes produced by the real world—comparing clouds to clouds.

Thus, the computational model does make a testable prediction. In fact, it makes even more testable predictions than the equilibrium model. For example, our model might predict that shifts in demand result in less variation in price than shifts in supply or that price changes greater than some percentage create a tipping phenomenon in which many people switch sellers. Kirman's (1997) used an agent-based model of the fish market at Marseille that predicted long-term buyer and seller relationship patterns that were evident in the data from the actual market.

Owing to the path dependencies, multiple equilibria, and even the absence of equilibria altogether in many agent-based models, testing can often be more difficult. Models that settle into equilibrium tend to include primarily negative feedbacks. When a firm makes positive profits, other firms enter and wipe out those profits. Here, actions are offset by other actions. In contrast, systems that generate complexity tend to include positive feedbacks as well. When one politician takes a new policy position, it creates incentives for other politicians to move as well, and when those politicians move, they set in motion an endless sequence of further movements. When one person commits a crime, they alter the incentives for others to do so, resulting in a cascade of crimes and victims.

Systems with mostly negative feedback tend to be very stable and predictable. Extraneous factors left out of the model can even be absorbed by the actions of the agents, leading to even less noise than we would expect from a prediction relying on the Central Limit Theorem. However, in systems with positive feedback, we loose some predictability. Small differences can build upon themselves and create large differences, making precise prediction difficult.

Empirical testing of complex systems models may require new advances in statistics. Many complex systems research efforts focus on analyzing the key temporal and spatial patterns that emerge in the model. While the human mind is quite adept at recognizing such patterns, we do not yet have a suite of easily applied statistical tools that can assist in this task. Such tools, along with advancing the study of complex systems, are also likely to have a variety of more direct empirical applications, such as recognizing useful patterns arising in real-world phenomena like markets and geographic-location decisions.

Even as purely abstract objects, computational models are useful. They provide an "artificial" reality in which researchers can experience new worlds in new ways. Such experiences excite the mind and lead to the development of novel and interesting ideas that result in new scientific advances.

5.3.6 Computational Models Are Hard to Understand

Finally, computational models are often dismissed because it may be difficult to fully understand the structure of the model and the various routines that drive it. Regardless of how the models are communicated, it is always easier to describe and understand simpler models. The actual computer code itself is a complete specification of the model, but there is a big difference between a complete specification and an accessible one. Indeed, most computer programmers have had the experience of looking at someone else's code (or even their own) and not being able to decipher it without a very intensive analysis.

Part of the issue here is that there are not commonly accepted (or understood) means designed to communicate computational models. There are, however, some notable efforts in other fields. For example, in object-oriented software design, after many separate efforts, a standardized way for specifying designs known as the Unified Modeling Language (UML) has emerged. Perhaps UML, or some other variant, will become the lingua franca of agent-based object modeling.

Ultimately, computational modelers must strive to create simple, easily communicated models. At the core of any computational model, like any mathematical model, there needs to be a simple set of driving propositions. It is these propositions that make up the model, not the apparatus that surrounds them. Core propositions in computational models are surrounded by lines of computer codes; in mathematical models, such propositions are surrounded by various solution techniques arising from, say, the calculus or linear algebra. In either case, it is the core propositions that we need to focus on and communicate.

5.4 New Directions

Computation as theory has an enormous potential for allowing us to investigate and understand better key phenomena, especially in complex adaptive social systems. While computation as theory can have many meanings, and surely over time it will acquire new ones, the focus of the models in this book is on computations that are designed to improve our theoretical understanding of the world by relying on agent-based objects.

Thus, we are more concerned with understanding the processes underlying the computations, as opposed to the computations themselves. To understand the underlying processes, we may well need to acquire direct experience with the computations through a series of carefully planned experiments, but the goal behind such work is the ability to create a set of propositions that apply both to the computational system *and* to other, more general, systems. Sometimes, running a single computation that shows the existence of some property of general interest could qualify under this definition, though computations that are used to predict the outcome of a particular (and often complicated) set of equations, rather than understanding the underlying processes, would not.

Second, we focus here on models that are composed of a set of simple algorithmic components, each associated with an individual agent. This requirement may seem a bit quirky at first—if our real concern is generating theoretical understanding from computation, why should we force these computations to use agent-based objects? Part of our reason for adopting this requirement is that it allows us to focus our efforts on an important, and relatively new, aspect of computation as theory.[4] More importantly, we feel that this requirement directs us into a realm of modeling that is likely to be very productive for understanding complex adaptive social systems. As discussed in the next chapter, these models have comparative advantages over other techniques (apart from convenience) that allow us to explore heretofore inaccessible problems that may hold the key to understanding more fully the behavior of complex adaptive social systems.

---

[4]There are many existing references, such as Judd (1998), that cover the numerical methods aspects of social science computation.

CHAPTER 6

Why Agent-Based Objects?

> Water which is too pure has no fish.
> —*Ts'ai Ken T'an*

> Scientific laws have conventionally been constructed in terms
> of a particular set of mathematical functions and constructs,
> and they have often been developed as much for their
> mathematical simplicity as for their capacity to model the
> salient features of a phenomenon.
> —*Stephen Wolfram, "Computer Software in Science and*
> *Mathematics"*

> With many calculations one can win; with few one cannot.
> —*Sun Tzu, The Art of War*

AGENT-BASED OBJECT models offer a new theoretical portal from which
to explore complex adaptive social systems. Like any theoretical tool,
these models have comparative advantages for certain types of explorations. In fact, their advantages appear particularly well suited, and
perhaps even necessary, for helping us to understand better the types
of problems that arise in the study of complex adaptive social systems.
Many of our existing tools tend to purify the theoretical waters so much
so that we are often left with a model that is barren of any useful signs
of social life. Tools like agent-based object models allow us to create new
theoretical ponds that can harbor simple, yet thriving, social ecosystems.

Much of the discussion in this chapter contrasts agent-based object
modeling with more traditional mathematical tools (see table 6.1). As
previously mentioned, no single theoretical tool is suitable for all needs,
and we are certainly not claiming that agent-based object modeling is
an exception. We do, however, suggest that the constellation of features
offered by such models represents a very appropriate set from which to
gain new insights into complex adaptive social systems.

6.1 FLEXIBILITY VERSUS PRECISION

An important feature of any theoretical tool is its trade-off between
flexibility and precision. Flexibility occurs when the model can capture a

TABLE 6.1
Modeling Potential

Traditional Tools	Agent-Based Objects
Precise	Flexible
Little process	Process oriented
Timeless	Timely
Optimizing	Adaptive
Static	Dynamic
1, 2, or ∞ agents	1, 2, ... , N agents
Vacuous	Spacey/networked
Homogeneous	Heterogeneous

wide class of behaviors; precision requires the elements of the model to be exactly defined.

One approach to achieving maximum flexibility is to use long verbal descriptions of the phenomena of interest. This tradition is well established in economics; for example, Smith's (1776) rather lengthy *Wealth of Nations* models a grand scheme of how the selfish behaviors of individuals can result in an outcome with surprisingly organized aggregate behavior. Such verbal descriptions, while flexible, often suffer from a lack of precision. There is an inherent ambiguity to such theorizing in terms of what is being expressed. More important, the implications of these types of descriptions are often difficult to verify—it is possible to make an apparently logical and coherent verbal argument that may in fact contain serious flaws.

At the other extreme are precise tools, like those embodied by some mathematical techniques. These techniques allow us to define a set of phenomena precisely and then solve the resulting system using a standard set of solution methods. Unfortunately, the cost of such precision is often a lack of flexibility in the phenomena that we can explore. To employ these solution methods, we need to have the components of the model pure enough so that they can be easily manipulated. While at times this added purity is an acceptable trade-off, it is easy to distill things so far that the system we are studying is of little interest or application.

Computational models represent an interesting trade-off between flexibility and precision. Computational models are remarkably flexible in their ability to capture a variety of behaviors. Many, and perhaps all, systems of interest to social scientists are likely to be computationally complete—that is, able to be encoded by a general purpose computer language. At the same time, computational models also require a high degree of precision. The computer program contains all the information about the assumptions of the model in a relatively compact form. Moreover, for the program to compile, there must be some level of logical consistency among the various parts of the program. This level of consistency is at a relatively low level: program statements must not directly contradict one another within the context of the required calculations.[1]

6.2 Process Oriented

By their very nature, computational models require a high degree of precision with respect to the underlying processes involved in the model. For a computational model to run, every aspect of how agents are allowed to interact must be well specified. Such issues are often ignored in mathematical models. For example, consider a simple market composed of equal numbers of buyers and sellers trying to trade with one another. A computational implementation of such a model requires us to define carefully when each agent is allowed to act, with whom it can interact, and its set of possible actions. Moreover, we must also specify what information each agent has access to, how it can use that information, how to resolve simultaneous offers, and so on. While any modeling method *could* incorporate such details, it is often the case that such process details are essentially ignored and taken care of by a narrow set of, often unconscious, defaults.

Appreciating the lack of process precision we employ in our modeling is difficult until one takes a standard model and implements it computationally. Indeed, it is worthwhile to program a "simple" market model or, for the less ambitious, to construct a simple algorithm for determining the competitive equilibrium "price" and "quantity" in a world with discrete values. Programming such models is often an enlightening experience in terms of the amount of process we ignore when we use traditional tools. Although some of the most interesting driving forces in social systems are related to process, being able to ignore things may be a real advantage in modeling (though, of course, explicitly knowing about and appreciating the impact of what we are ignoring is still needed).

Given both the flexibility and precision inherent in computational models, these methods can be a nice way to structure new problems. Computational implementations of problems often illuminate the key features and processes that must be modeled. The well-defined nature of computation can often point us toward new frameworks from which to make theoretical progress. Much of the creativity required for

---

[1] Thus, there are no guarantees that the program is logical given the goals of the model.

developing such models is in finding a good way to represent the key issues (in computational learning, this is known as the "representation problem").

Traditional models of technological progress in economics often rely on a single parameter in a production function to represent "technology." While implementing such a parameter is a nice way to simplify the world, it is so drastic a simplification that many of the important issues we normally associate with technology get lost. In computational work, a natural way to model technology is as a bit string, where each bit represents a different binary technological choice (for example, whether to have two or four doors on a car or to add tail fins). We can then make the value of a particular set of technological choices depend on some, perhaps nonlinear, function of the bits. Given this representation, we can begin to structure some potentially interesting questions; for example, search can become a cumulative process of discovery, technological breakthroughs can be modeled as "mutations" in the bit string, and more elaborate innovations, such as when one firm appropriates the ideas of another, can be captured via larger "swaps" among bit strings.

6.3 ADAPTIVE AGENTS

A long-standing interest of social scientists is how bounds on the ability of agents to rationally process information impact the behavior of social systems. A related set of questions concerns the influence of learning on such systems. At one extreme, is the belief expressed by Friedman (1953) and others that evolutionary mechanisms result in systems composed of only those agents who employ high degrees of rationality and information-processing ability. Thus, assuming that outfielders in a baseball game have the ability to manipulate Newton's equations rapidly and determine the exact spot on the field where a pop fly will land may be sufficient to predict where the outfielder will end up.[2] (Friedman used the example of billiards, but we choose baseball for reasons that will become obvious later.)

A priori, it is not clear that evolution must lead to optimization. Evolutionary systems often get stuck at local optima (for example, many organisms eat and breath through the same tube, even though this often causes them to choke). It may be that adaptive social systems act more like two campers fleeing from a marauding bear, where the goal of each camper is not so much to outrun the bear as it is to outrun each other.

---

[2] Anyone who has played on an academic department's softball team may have good reason to doubt even this assumption.

Clearly, the evolution-leads-to-perfection argument is one that is worthy of testing directly, as knowing the exact conditions under which it holds would allow us to apply our other theoretical tools better.

The flexibility of computational tools make them well suited for considering models of boundedly rational agents who adapt their behavior. In fact, computational models of learning have been developed in a variety of fields, including computer science, physics, and psychology. Early work in this area focused on relatively high-level cognitive models that solve problems by manipulating symbols. The initial success of these efforts on toy problems was quite striking and suggested to researchers that within a very short time such models would form the basis for, say, world-class-level chess programs. Real problems proved much more difficult than initially thought. For example, IBM's Deep Blue program beat Kasparov not by matching the elegance of Kasparov's 2- or 3-position evaluations per second, but rather through a relatively crude brute force algorithm capable of evaluating 200 million positions per second.

An alternative to the symbolic approach to learning is a class of low-level adaptive algorithms, such as genetic algorithms and parallel distributed processing. This alternative approach was initially criticized as being too unstructured and, regardless, not needed given the promise of the more top-down cognitive approaches. The low-level adaptive algorithms, however, gained ground when the limits of the top-down approach become more apparent.

The ability to analyze systems of "adaptive" agents systematically is an area of great promise for social scientists, but it does face a potentially serious scientific challenge: can we create a coherent science of adaptive agents? One advantage of optimization-based models is that there is typically only one way for an agent to be optimal while, as we all have experienced at one time or another, there appears to be an infinity of ways for an agent to be "dumb." Thus, we could find ourselves in a situation where an ever growing zoo of adaptive agents arises and the field quickly becomes mired in endless debates about the appropriate way for agents to be dumb. One resolution to this potential quandary would be to agree on a particular set of adaptive assumptions, perhaps based on experimental data on human learning or some other criterion. A more interesting and, in our opinion, more promising approach is to "let a thousand flowers bloom" in hopes that large equivalence classes of adaptive behavior will be discovered. A number of computational experiments already suggest that seemingly different adaptive algorithms behave in very similar ways.

The quest for understanding better the features of adaptation that lead to common behavior is a key scientific question. At the moment, the evidence from computational models hints at the potential existence of a large equivalence class of adaptive behavior. If such a hypothesis is confirmed, we will be able to model more easily the behavior of adaptive systems, as the exact implementation of details will not matter. More important, such a result would help us unify our understanding of a variety of systems, both natural and artificial.

6.4 INHERENTLY DYNAMIC

Many of our existing analytic tools avoid an emphasis on dynamic processes and focus on equilibrium states. When transition paths are short and conditions are stable, such an approach may provide a good description of the world. In natural systems, however, equilibria are usually associated with the death of the system. The conditions that favor equilibrium analysis are likely the exception rather than the rule in many complex adaptive social systems. If so, the techniques that we traditionally use to analyze these systems may be like trying to "understand running water by catching it in a bucket."

Even when the conditions are right for equilibrium analysis, understanding the dynamics of the system may still be important. In models with multiple equilibria (a situation that is often intentionally avoided by theorists), dynamic considerations may be used to select among equilibria. In adaptive systems with multiple equilibria, we often find that certain equilibria are associated with larger basins of attraction[3] and thus are much more likely to trap the adaptive agents. Similarly, the stability of a particular equilibrium is tied closely to the dynamic behavior of the system. Finally, dynamic notions can be used to clarify the transition path and time to equilibrium. While a proof that the system will, say, asymptotically converge on a particular equilibrium is very useful, the importance of that result depends on whether the transit time is a few, or a few billion, iterations.

A nice example of where the dynamics are interesting even when a single, well-defined equilibrium exists is the previously mentioned case of catching a baseball. While it is true that the equilibrium analysis of an optimizing, Newtonian outfielder does make an accurate prediction of the player's equilibrium position on the field, this model does quite poorly at predicting the actual path to that point. Outfielders do not run in a straight line from whatever their location happens to be when the ball is hit to the place on the field where the ball will land. Instead,

---

[3] Basins of attraction are the areas of the space where the dynamics lead the system to a common outcome. Thus, they are similar to watersheds—regions of land where water drains to a common outlet—in the physical world.

studies (for example, McBeath et al., 1995) indicate that outfielders run in an arc-shaped path that is consistent with a simple, vision-based behavioral heuristic that keeps the ball on a linear trajectory relative to the background. Depending on what we are trying to predict or understand, this difference in behavior may or may not be important. For example, if outfielders were not allowed to move but got paid by predicting where the ball would land based on the initial information, it is doubtful that our equilibrium model would be of much use.

A more mathematical example of some of these issues comes from a Markov model of, say, strategic nuclear armament. It can be shown that this model has only a single absorbing state, namely a world that is completely destroyed by nuclear war. Thus, the model predicts that we will end up (with probability one) in such a state. Obviously, notwithstanding the strong prediction, most of the interest in this kind of system is in its (hopefully very long) transient behavior.

Social scientists have often recognized the importance of dynamic analysis but have been very constrained by their tools. According to Von Neumann and Morgenstern (1944, 44), in their seminal work on game theory, "We repeat most emphatically that our theory is thoroughly static. A dynamic theory would unquestionably be more complete and therefore preferable. But there is ample evidence from other branches of science that it is futile to try and build one as long as the static side is not thoroughly understood."

Computational models using agent-based objects are a very natural way to explore the dynamic behavior of a system. Regardless of the presence of equilibria, such behavior is often the most interesting part of the system. As Ursula Le Guin (1969, 220) said, "It is good to have an end to journey toward; but it is the journey that matters, in the end." In situations in which equilibria are a possibility, understanding the dynamics is likely to be insightful. In situations where equilibria are nonexistent or transient paths are long, understanding the dynamics is critical.

6.5 Heterogeneous Agents and Asymmetry

Most of our existing analytic tools require that the underlying agents have a high degree of homogeneity. This homogeneity is not a feature we often observe in the world but rather a necessity imposed on us by our modeling techniques. Unlike traditional tools, computational methods are able to incorporate heterogeneous agents easily.

Whether we actually need to model heterogeneity is an important research question. It may be the case that given sufficient agent heterogeneity, the aggregate behavior of the system may no longer depend on the various details of each agent, and abstracting this behavior into a single representative agent is feasible. This is an open question, and we can use computational models to address it directly. All modeling, at some point, needs representative agents to emerge. The key is to make sure that we get the right representatives for the right reasons.

A second area in which we are forced to simplify our models is in the amount of asymmetry we have in the system. Like homogeneity, symmetry assumptions dramatically simplify calculations, and so they are used even though asymmetry may be a pervasive and influential feature of social systems. We know from some mathematical models that simple asymmetries in, say, information can alter our prediction of a single, well-behaved equilibrium point to one where there are multiple equilibria linked to, say, agent expectations. Computational models that use agent-based objects can easily accommodate asymmetries.

6.6 SCALABILITY

The ability to solve a model analytically is often tied to the number of agents that are used. Thus, traditional methods typically focus on models composed of either very few or very many agents. In physics, for example, mathematical methods exist for modeling planetary motion with two, three, and an infinity of planets. The intermediate cases are too difficult to solve analytically and must be solved computationally. In economics, we have good mathematical models of industrial behavior with monopolies, duopolies, and perfect competition. Once we begin to analyze systems of oligopolies, however, we are confronted with a lot of theoretical ambiguity.

Models with agent-based objects are easily scaled. Once the behavior of a single agent is described, it is usually easy to explore the behavior of systems of essentially arbitrary size by simply adding more agents to the system.

There are many examples where scaling up a system even slightly can have dramatic effects. In economics, we see such changes when we move from one to two to three firms in our models of industrial behavior. Adding a new dimension to a system can often cause its behavior to change dramatically as well. There are examples, such as in percolation and spatial voting theory, where a theorem becomes impossible to prove (or even becomes contradicted) as we add another dimension to the problem.

Being able to manipulate easily the scaling of our models may promote the discovery of key scaling laws for complex adaptive systems. In biology, the branching features of a variety of respiratory systems scale as a fixed power of body size across at least twenty-five orders of magnitude. In economics, city and firm size distributions appear to follow well-known scaling relations. Computational models may offer a suitable Petri dish from which to initiate a more direct investigation of social scaling laws.

6.7 REPEATABLE AND RECOVERABLE

Computational models provide some unique opportunities as an experimental medium. A lot of theory is inspired by anomalies observed in the real world, and the artificial worlds created by agent-based objects can provide similar inspiration. Unlike the real world, however, anytime we observe an anomaly in a computational model the initial state of the system can be recovered, and we can "rerun the tape" and observe the old system from whatever new perspective is needed to reveal the cause of the anomaly. This ability to rerun and reprobe a system facilitates the rapid development and refinement of theoretical ideas.

Computational worlds are also repeatable, allowing multiple observations on the "identical" system. Not only can these systems be intensively probed, as we have discussed, but subtle experiments can be conducted with a degree of precision unattainable in real experimental settings. In real experiments, especially with human agents who often alter their behavior based on experiences or expectations, it is impossible to repeat an experiment with the same subject under near-identical conditions. In computational models it is easy to take back the experiences of the subjects and run them anew with slight alterations in the parameters. Moreover, other elements that often confound experimenters, such as manipulating payoffs, expectations, and risk aversion, can be tightly controlled in artificial worlds.

6.8 CONSTRUCTIVE

Agent-based object models inherently provide constructive "proofs" to propositions. In particular, once we specify an agent-based object model and find that it leads to a coherent macrophenomenon, we have thereby found at least one set of microconditions that is sufficient to generate the macro-observations. This, of course, does not imply that our set of conditions is the one that actually produced the phenomenon in the real world. Nonetheless, Epstein (1999) argues that this "generative" approach—that is, we must grow it to show it—is a distinct and powerful way to do social science.

The ability to fully generate a phenomenon from the bottom up often provides new avenues of insight and understanding. Of course, science can make, and has made, great progress without constructive explanations, but this more often than not reflects the difficulty of attaining such explanations rather than their desirability. Having a complete view of a phenomenon, from its origins to its manifestation, is inherently satisfying. While we can, say, do biology by not worrying about how DNA came into being, the origins of life address a deep scientific curiosity. More important, by understanding the origins of a phenomenon we often gain new insights into its current manifestations, even when the generative process is no longer active.

By analogy, consider proving a proposition using a proof by construction versus one by contradiction. Proofs by contradiction are a perfectly legitimate way to establish the validity of a proposition. Nonetheless, they typically are not that intellectually satisfying—at some point during the proof a contradiction is established and therefore the original proposition must hold, but the contradiction rarely provides any useful insights into the underlying problem beyond proving the proposition at hand. Unlike proofs by contradiction, the formulation of a constructive proof often provides new avenues from which to venture forth with new propositions.

Hayek (1945, 530) made a direct appeal to generative explanations with respect to market phenomena:

> The problem is thus in no way solved if we can show that all the facts, *if* they were known to a single mind (as we hypothetically assume them to be given to the observing economist), would uniquely determine the solution; instead we must show how a solution is produced by the interactions of people each of whom possesses only partial knowledge. To assume all the knowledge to be given to a single mind in the same manner in which we assume it to be given to us as the explaining economists is to assume the problem away and to disregard everything that is important and significant in the real world.

6.9 Low Cost

Sometimes computation may be necessary to solve theoretical questions; other times it is merely convenient. Computational methods tend to be very cost effective. While developing the initial computational model can be costly, the marginal cost of running or modifying it is usually very low. Thus, once the model is developed, it is easy to run sufficient trials to accommodate any statistical necessities or to incorporate additional factors in the experimental design.

There is a simple economic argument justifying the use of computational models. Over the past few decades, computational costs, in terms of hardware, software, and human capital, have experienced significant declines. During the same period, traditional analytic methods are starting to operate under conditions of high diminishing marginal returns. Thus, the costs of acquiring new theoretical results using traditional means keep rising, while the costs of doing so using computational methods remain on a favorable part of the production function. Given these differences, producers of theoretical ideas should begin to substitute computational methods for more traditional ones.

6.10 ECONOMIC *E. coli* (*E. coni*?)

The interaction between the theorist and the computational model provides an ideal medium from which theoretical insights can be gleaned (Tesfatsion, 1997, 2006). Agent-based object models give the theorist some rather intimate experiences with the phenomena of interest. As we have outlined, these artificial worlds are fully observable, recoverable, and repeatable, and thus they are a fertile playground from which theories can be created, refined, and tested. Like many theoretical tools, computational models have the potential to produce insights well beyond those needed to implement the original model.

A fanciful, but perhaps ultimately enlightening, use of agent-based object models is as an "animal" model for the social sciences. The ability to experiment with animal models like *E. coli* in biology and *Drosophila* in genetics, has led to great advances in our understanding of human systems. Unfortunately, there is not an obvious choice of an animal model for social systems research. Indeed, even human-based experiments are relatively new in fields like economics, where their results are just beginning to facilitate the process of scientific creative destruction. While the possession of a simple animal model is not necessary for scientific progress—both economics and astronomy were developed around passive-observation-based methodologies—having an animal model may lead to new scientific opportunities.

Computational models composed of artificial adaptive agents (with apologies to Linnaeus, we could call the critters *E. coni*) could prove to be a productive way to develop new social theories. These simplified economic systems might be just sufficiently complex to allow "real" economic behavior to emerge in a more observable and understandable world. Computational Petri dishes of *E. coni* could be used to create social equivalents of the Galapagos Islands and, in so doing, help us attack some central questions: How general are our theories? How much agent sophistication do we need in a system before it becomes social? At what point do agent details cease to be important?

Such "alternative" worlds offer wonderful opportunities to even those theorists wedded to traditional tools. For example, artificial-life models like Packard's (1988a) work using artificial bugs seeking food on a landscape or Epstein and Axtell's (1996) Sugarscape model are often dismissed by economic theorists as lacking useful economic content. Such arguments seem misplaced (especially given the usual lack of timidity in the application of economic theory to other areas). At some, albeit simplified, level these systems are all about the processes that economists hold most dear: scarcity, choice, and exchange. As such, they can serve as a nice test bed for our theories. How well do standard theories, perhaps developed to explain more sophisticated phenomena, operate in these simplified worlds?

While creating artificial worlds of *E. coni* may at first appear to be a bit odd, the actual cost of doing so is low and the potential benefits are high. Indeed, we may already have such worlds emerging in the form of software agents, either real or artificial, interacting on the Internet.

Our theories are often developed based on, to paraphrase Langton (1989), social life as we know it rather than social life as it could be, and having access to some alternative universes to explore should allow us to develop better theories. Suppose we happen upon an alien civilization, what a priori predictions would we make about its social and economic behavior?

PART IV

Models of Complex Adaptive Social Systems

In the chapters that follow, we explore a variety of models of complex adaptive social systems. We have a few goals for these explorations. First, and foremost, we want to uncover key insights into the behavior of complex adaptive social systems. We want to understand the behavior of both the agents within the systems and of the systems themselves. Second, we want our models to be as simple and accessible as possible. While simplicity is always a goal of modeling, here we are willing to err on the side of too much simplicity if it makes the resulting models more accessible to readers. At times, we forgo some research possibilities that, while interesting to specialists, would tend to obscure the work to others. One advantage of seeking such a deep simplicity is that it encourages coherence across all of the models and, in so doing, may allow a more general picture of these types of systems to emerge. Finally, in the models that follow we rely on a variety of analytic techniques (including computational modeling, mathematical proofs, and thought experiments) both to enhance our understanding and to demonstrate a range of analytic possibilities.

CHAPTER 7

A Basic Framework

> Now what, monks, is the Noble Eightfold Path? [It is] as
> follows: right view, right intention, right speech, right action,
> right livelihood, right effort, right mindfulness, right
> concentration.
>
> —*Magga-vibhanga Sutta*
>
> A foolish consistency is the hobgoblin of little minds, adored
> by little statesmen and philosophers and divines.
> —*Ralph Waldo Emerson, Self-Reliance*

COMPLEX ADAPTIVE SOCIAL SYSTEMS are composed of interacting,
thoughtful (but perhaps not brilliant) agents. Given this underlying
structure, models of these systems, especially those that rely on agent-
based objects, tend to confront a common set of issues. In this chapter,
we discuss some of these issues in hopes of illuminating the core modeling
elements and building some overall coherence. We make no presumptions
that the modeling paths we suggest here are inherently superior to other
possible approaches, and we suspect that there are likely to be many
productive alternatives.[1]

Given our focus on interacting systems of agents, it would be nice to
have a simple framework from which to discuss such systems. As we
show here, what appears at first to be a simple task is fraught with
difficulty. Indeed, it is easy to have a beautiful eightfold way collapse
into nothingness. That being said, such a destructive transformation is
insightful and we will proceed apace.

7.1 THE EIGHTFOLD WAY

Our task is to think about how to classify agent models. With the obvious
apologies in advance, the Noble Eightfold Path from Buddhism may not
be a bad place to start. The elements of the Eightfold Path can be mapped

---

[1] The study of complex systems is often conflated with the art of constructing agent-based
models, and while agent-based models are a valuable tool for understanding complexity,
other tools like mathematics and thought experiments are also needed in this quest.

TABLE 7.1
An Eightfold Mapping of Agent-Based Object Models

Path	Focus
View	Information and connections
Intention	Goals
Speech	Communication among the agents
Action	Interaction
Livelihood	Payoffs
Effort	Strategies and actions
Mindfulness	Cognition
Concentration	Model focus and heterogeneity

to key modeling issues in complex social systems (see table 7.1). As in any metaphorical mapping, sometimes the alignments prove tight, and other times we must resort to broad interpretations to maintain relevance.

7.1.1 Right View

Right View encompasses the information that an agent receives from the world. Such information can influence agents in both direct and indirect ways. Directly, incoming information will often cause an agent to immediately react to what was received by taking some action. Indirectly, information is often "memorized" via some change in an agent's internal state, and such changes may set the stage for actions that will only become realized far into the future.

Much of the potentially interesting science of view is in its infancy. For example, as Herb Simon pointed out, agents typically confront a wealth of information, and thus the scarce resource here is not information but rather attention. Given the inherent limits of information processing, agents must actively ignore most of the potential information that they encounter. We are typically not aware of how much information we ignore, though on occasion it does becomes apparent; for example, when we drive a car into an intense rainstorm, the amount of information available to us is often dramatically curtailed, yet we typically are able to continue to drive onward with only limited compromises on speed and safety. It may even be the case that agents operate more effectively with less information. We suspect that a full analysis of how agents selectively attend to information will provide some interesting scientific opportunities.

Agents typically supplement information garnered from outside with internally generated information. For example, agents may develop ways, such as statistics, to summarize the flow of incoming information so that it is easily stored and used. Alternatively, agents may generate elaborate internal models that allow them to transcend inherent perceptual limitations. For example, internal models allow you to visualize what is currently behind you, even though you have had very little recent visual input about that scene; or to recognize that an object, even when it has been obscured by a curtain and cannot be seen, still exists (a skill that develops at around nine months of age in humans).[2] Another use of internally generated information is to produce "would-be" worlds that may become important in the future. It is likely that there is an optimal amount of such "daydreaming"—too little and not enough information about what could happen will be generated, and too much and poorly anchored fantasies will begin to run amok.

A further complication is that the inputs that agents receive often come from other agents. As such, agents may be able to manipulate, at least partially, their outputs so as to influence the actions of others. As we will see, models where such manipulation is possible can lead to very interesting behaviors.

Networks may also be important in terms of view. Many models assume that agents are bunched together on the head of a pin, whereas the reality is that most agents exist within a topology of connections to other agents, and such connections may have an important influence on behavior.

In many of the models that follow, we make fairly direct assumptions about how the information flows and is perceived by the agents. The timing of the information flow can be important—what agents know and when do they know it can make a big difference to the outcome of a social process, and some of the models that follow explore this issue explicitly.

7.1.2 Right Intention

Right Intention focuses on the goals of the agents. In some models, agents are assumed to have a set of explicit, well-defined goals that direct action. In other models, goals are built in implicitly, such as in models of biology in which agents survive and reproduce only if they are able to acquire sufficient resources from the environment.

By manipulating an agent's intention, we can obviously put in place strong forces on the model's behavior. Of course, the most interesting results come about when the outcome of the model is, at some level, at odds with the induced motivations of the agents—to use Schelling's

---

[2]Magicians artfully manipulate internal models designed to assist perceptual tasks by "misdirecting" observers using cues as simple as the pointing of a finger.

terms, when the micromotives and macrobehavior fail to align. Thus, it is far more interesting to see cooperative behavior emerge when the agents are self-interested than when the agents are presumed to be altruistic, or to see agents aggregate into cities when their goal is to be left alone.

Along with agent intentions, we often have desires for the system as a whole. In social systems, we may want the agent behavior to aggregate in such a way that the system achieves some goal. For example, we may want market trade to result in efficient outcomes, political systems to create socially productive and fair policies, and so on.

7.1.3 Right Speech

Right Speech accounts for the information that agents send to others. Agents can send information to other agents by taking observable actions or, more explicitly, by using some communication channel. Models can differ in terms of the kind of information that is allowed to be communicated, how that information is allowed to flow among the other agents, and the quality of the information.

We know that communication is an important feature of all kinds of complex adaptive social systems, ranging from molecular signals that control cellular functions to the wording of international treaties that control global relationships among nations. The foundational work of Shannon (1948) in information theory provided a sound basis from which to think about very primitive issues in communication, yet these results only begin to provide insight into the actual complexities and applications of communication in the world around us.

Agent-based models offer some interesting opportunities to explore the notion of communication as a way to organize complex systems. Using such models, we can begin to explore "strategic" communication in which agents must decide what to say to others and how to react to what others say to them.

7.1.4 Right Action

Right Action embodies all of the interactions that occur among the agents. Each agent receives and processes information and, by its action (or even inaction), generates information that influences the other agents and the system itself. Such interactions depend on the "space" within which the agents are contained. This space could be defined by physical realities, like having the agents arrayed around the perimeter of a circle or by more abstract entities such as "friendship." Whatever the form of the space, it can mediate agent interactions by constraining the flow of information and action, such as when we only allow agents to interact within some well-defined neighborhood. Space is often endogenous in a system; for example, agents may prefer to interact with "friends," and over time the interaction possibilities may change as previous interactions alter the space of friends.

In some models, agents are assumed to gather input, process it, and act, simultaneously. In these models, some external synchronization device must coordinate the behavior of the agents. The amount of synchronization in real social systems probably varies dramatically. There are some external and normative mechanisms that do coordinate individual behavior. Thus, each morning individuals awake and choose clothing for the day, and this choice is more or less simultaneously revealed to others throughout the day. Social functions, like pot luck dinners or bringing gifts to a birthday party, have a similar flavor. Social systems also embrace institutions that coordinate behavior, such as sealed-bid auctions with fixed clearing dates or national election days.

Agents can also activate asynchronously. Under asynchronous activation, each agent awakes at a different time, processes whatever information is currently available, and then by its action alters the informational ether that will face the other agents when they are activated. Asynchronous activation requires that the agents be placed in some activation order. One way to order agents for activation is randomly (either with or without replacement); alternatively, agents could be ordered by some exogenous or endogenous characteristic. Thus, agents could be activated by, say, their location in physical space, such as we see when, say, canvassers go door-to-door seeking signatures for a petition, or by some other characteristic like age or seniority. Alternatively, we can activate agents based on incentives (Page, 1997). Under this mechanism, agents activate when they have an incentive to do so, for example, when the quality of their current situation deteriorates past some preset threshold or the value of changing exceeds some cost.

Experiments with different systems indicate that the type of updating can make a big difference to the outcome of the model, so the choice must be well informed. Sometimes, the choice can be guided by the realities of the actual system we wish to model. At other times, when the choice is more amorphous, alternative mechanisms need to be applied so that the dependencies of the model can be clearly identified.

7.1.5 Right Livelihood

Right Livelihood concerns the payoffs that accrue to the agents. Payoffs can arise via the pure "physics" of the model, where actions aggregate to change the world in such a way that the resulting outcomes provide some benefit to, or impose some cost on, the individual agents. Of course, agents also have the ability to change the physics of the world by co-creating new opportunities for payoffs—for example, when they make a zero-sum bet on some outcome.

Payoffs can play many roles in these types of models. By assuming that agents have the goal of improving their payoffs, modelers can impose a lot of structure on the behavioral possibilities of the agents. We may also want to use payoffs as a way to drive adaptation by letting agents reproduce based on their performance. Payoffs can also be used to determine the activation order for agent updating.

7.1.6 Right Effort

Right Effort embraces agent strategies and actions. One feature that makes social science particularly interesting, and difficult, is the way in which agents anticipate and react to the potential behavior of other agents. Strategies can take many forms, from simple-fixed heuristics to elaborate optimization routines that change over time. Even these types of categories are not always well defined. For example, the fixed rule being employed by an agent to guide strategic behavior may be the result of an elaborate and complicated optimization procedure.

In many contexts, people do appear to be following rules (see, for example, Camerer, 2003). While the idea that we do the best we can given our constraints has great intuitive appeal, it is equally compelling to think of individuals as relying on simple to moderately involved heuristics that tend, in general, to result in "good" outcomes. Such heuristics may go wrong at times, and indeed a lot of work in the area of behavioral decision theory is focused on finding situations where normally useful rules go bad. After all, to err is economic.

There are many reasons why people may exhibit less behavioral plasticity, and therefore less sophistication, than is commonly assumed by rational choice theorists. First, as long as a rule satisfices, agents may not see the need to change. Second, people may lack the ability to infer causal relationships between actions and outcomes. Causal inferences become increasingly more difficult the more complex the environment or the less exposure an individual has to the particular decision scenario. Finally, if events transpire quickly, such as in a standing ovation or a riot, there is just not sufficient time to contemplate an optimal strategy and agents may follow existing, or make up new, behavioral rules.

While we believe that rule-following behavior is an important element of human systems, there is ample middle ground. Humans have the capability to be extremely thoughtful and careful in their decision making, and it may be the case that on some decisions, perhaps investing money or buying a house, decision making is consistent with a utility maximization framework. However, in other choice contexts like proposing marriage or participating in revolutions, humans may rely more on intuitions, emotions, and gut instincts.[3]

Within the set of possible rules that agents might use, some rules are simpler than others. Rules like those used in, say, Conway's Game of Life (see Gardner, 1970) are easy to understand and execute, while those in, say, Samuel's (1959) checkers-playing program are much harder to comprehend, even though they both can run on the same computer. Of course, ease of execution is not a good metric for the amount of intelligence a rule embodies, as brilliance is often sublimely simple. For example, the careful optimization and recursive thinking that lie at the heart of much of game theory often result in strategies that take the form of very simple behavioral rules. To illustrate this point, consider participating in a second-price (also known as a Vickery) auction in which a good is sold to the highest bidder at a price equal to the second-highest bid. The optimal strategy in such an auction is to bid your true value for the good—a very simple rule, but one that arises only after some clever strategic thinking.

More generally, in a reasonable class of games we can construct a best-response correspondence, $BR(X)$, that maps the actions of the other agents, X, to a best action. Such a correspondence provides a fixed rule—if my opponents do X, then I should do $BR(X)$. Thus, fixed-rule-following agents may arise from hyperrational strategic thinking.[4] Indeed, if we allow agents to have arbitrary beliefs over what others might be doing, then it becomes possible for almost any behavior, including fixed rules, to be recast as optimal given some set of beliefs.[5]

Thus we see how at the extremes of behavior—ranging from mindless rule-following to sophisticated optimization—agents embrace fixed rules. Agents that follow rules may do so either because they are simple or transcendent. It is when we move in between these two extremes that we find ourselves in a world of messy, and perhaps even sophisticated, computations. For example, models of learning from psychology (such as Hebbian learning [Hebb, 1949] and neural networks [Hopfield, 1982; Kuan and White, 1994]) and economics (like quantal response updating [McKelvey and Palfrey, 1998] and experienced weighted utility [Camerer and Ho, 1999]) rely on many moving parts and substantial

---

[3] Alas, even the notion of "gut instincts" is not as simple as it appears. In vertebrates the gut is controlled by the enteric nervous system, which is relatively localized and autonomous, and appears to rival the spinal cord in terms of complexity and function.

[4] As we know from the Game of Life, a system composed of fixed-rule-following agents does not guarantee the existence of an equilibrium. Even if equilibria exist, convergence depends on the response functions being smooth and not overly steep near the equilibria.

[5] Ledyard (1986) proves a related result for Bayesian equilibrium.

computations. This suggests an imperfect correlation between the intelligence and strategic sophistication of an agent and the observable level of computation employed by that agent. It also implies that we may need to enter a messy domain if an accurate representation of agent behavior is needed to model the world successfully.

7.1.7 Right Mindfulness

Right Mindfulness is the level of cognition employed by an agent: how smart should agents be? It is true that most agent-based models rely on simplistic agents, and people are often more sophisticated. Of course, as the evidence from behavioral economics mounts, it also appears that people are often less sophisticated than most game theory models assume. More likely than not, the sophistication of the agents is context dependent, and in some situations attempts at optimization predominate, while in others simple heuristics are employed. Indeed, agents may vary their cognitive commitment to the task at hand, and such heterogeneity may be an important driving force in the world. The important question is not whether agents are boundedly rational per se, but rather when and how does this make a difference.

The mindfulness of social agents differentiates them from physical agents. Social agents often have mental models that they use to inform their behavior. Moreover, unlike physical agents, there is a plasticity in social agents who can change how they behave if outcomes are not to their liking. The rates and mechanisms of change may well depend on the system. An individual human agent can conduct a "thought" experiment and rapidly alter her behavior, while a lower-level biological agent may be destined for much slower changes via less direct mechanisms like natural selection. In contrast, bosons, quarks, electrons, and atoms, at least as far as we know, cannot change their rules modulo quantum fluctuations.

Ultimately, there appears to be no context-free answer to the question of how smart should we make our agents. Complex social systems models do—and should—vary in the level of sophistication embedded in the agents. There are models like the Game of Life where agents are given no ability to think or strategize, and the scientific exploration surrounds understanding how these rules result in productive, macrolevel properties—in short, how the simple can create the complex. There is another branch of work, for example the Double Auction Tournament (Rust, Miller, and Palmer, 1992, 1994) and the annual Trading Agent Competition (Wellman et al., 2003), where extremely sophisticated agents, often designed by teams of scientists over months or years to embody either actual or idealized human behavior, lie at the core of the analysis. This work seeks to find the regularities arising from such systems—how the complex can create the simple.

Recall that there is no requirement that the elements of a model match the elements of the system being investigated. That being said, agent-based computational models do allow us to create agents that can begin to emulate the behavior of humans. Of course, even this goal is problematic as it is doubtful that we can fully specify the "behavior" of a human. One problem is that it is often difficult to glean strategic behavior from simple observations. For example, a person cooperating in the Prisoner's Dilemma game might unthinkingly do so because she just wants to be nice or might be a highly intelligent and strategic thinker optimizing self-interest based on her reading of Kreps et al. (1982). Even the notion of *a* level of human rationality is suspect, as the amount of cognitive attention and commitment likely varies across individuals and even within one individual across contexts. There is a long tradition in economic modeling that assumes that people are identical in their cognition but vary in their preferences. Perhaps the next advance is assuming variation in thinking across individuals (or choice domains).

7.1.8 Right Concentration

Right Concentration is the focus of the model—namely, it requires the model to be just sufficient to capture the phenomenon of interest. Models always have contexts, and what works well in one context may fail in another. If we want to understand the essence of cooperation, then perhaps we ignore network topologies. If we want to understand the importance of connections, then perhaps we should simplify the domain of action.

Another element of concentration is the amount of heterogeneity in the model. Within a given system there can be substantial heterogeneity across agents. Many economic models investigate worlds of multiple agents via a single, "representative" agent. Thus, while there are many agents in the model, they are all identical. The advent of agent-based methods allows the investigation of populations of truly heterogeneous agents. Heterogeneity enters these models in various ways. One method is to have an "ecology" of agent types, each relying on different behavioral governing mechanisms. Alternatively, one can use homogeneous agents but allow differences in histories, information, or underlying characteristics to cause behavioral differences among the agents.

Models of complex systems phenomena should be simple, not complicated. This point often seems to get confused and twisted in various ways, but the point of modeling—whatever the target—is to simplify an otherwise overly complex world. Thus, even when the resulting behavior is complex, the underlying model should be simple.

7.2 SMOKE AND MIRRORS: THE FOREST FIRE MODEL

To clarify some of the ideas embedded in the Eightfold Way, we introduce a simple model of forest fire dynamics. This model provides a nice example from which to explore various topics, as it is stark enough to be explained easily yet results in behavior that is subtle enough to be interesting. Moreover, the model serves as a convenient springboard from which to explore not only complex systems, but complex adaptive ones as well. The discussion proceeds by introducing a basic model assuming fixed, homogeneous rules and from there further developing the model by introducing additional layers of agent sophistication.

7.2.1 A Simple Model of Forest Fires

Consider a world in which trees grow along a line known as Thunder Ridge. Each spot on the ridge is suitable for growing a tree. Each spring there is a fixed probability, g, of a tree sprouting up in an unoccupied spot. To keep things simple, once sprouted, trees immediately grow to their full size and remain that way unless disturbed. In the summer, lightning storms hit the ridge. Each spot on the ridge has a probability f of being struck by lightning. If a tree gets struck, it catches fire and the conflagration spreads to all contiguous trees. Empty locations act as fire breaks, preventing the further spread of the fire.

Table 7.2 illustrates the Forest Fire model. At time period 1, the forest is empty. During the growth phase (1.G), trees spontaneously arise on the ridge in locations designated by "t." During the lightning season, some trees are struck and set ablaze; thus at time 1.F two trees (designated by "T") have been hit. At the start of period 2, the struck trees and any connected neighbors have been burned to the ground. This cycle of growth and fire continues over subsequent time periods.

7.2.2 Fixed, Homogeneous Rules

As formulated, this agent-based model is very simple. The agents are the individual locations. Each agent follows an identical, fixed rule. In the spring, if your state is currently empty, then change your state to having a tree with probability g; otherwise maintain your current state. In the summer, if you are in a tree state, with probability f catch fire

TABLE 7.2
A Simple Forest Fire Model

Time	Forest																			
1	-	-	-	-	-	-	-	-	-	-	-	-	-	-	-	-	-	-	-	-
1.G	-	t	-	t	t	-	t	t	-	t	t	-	-	t	-	-	-	-	t	t
1.F	-	t	-	t	t	-	t	t	-	t	T	-	-	T	-	-	-	-	t	t
2	-	t	-	t	t	-	t	t	-	-	-	-	-	-	-	-	-	-	t	t
2.G	-	t	t	t	t	-	t	t	-	-	-	t	t	-	-	t	-	-	t	t
2.F	-	t	T	t	t	-	t	t	-	-	-	t	t	-	-	T	-	-	t	t
3	-	-	-	-	-	-	t	t	-	-	-	t	t	-	-	-	-	-	t	t

Note: A t indicates a newly grown or existing tree and a T designates a tree struck by lightning.

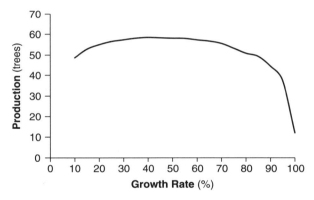

Figure 7.1. Tree production in a Forest Fire model with $f = 0.02$ and homogeneous, fixed rules.

or catch fire if any contiguous trees have caught on fire. Once an agent catches on fire, it reverts to the empty state.

Despite its simplicity, this model yields some provocative results. Define production of the forest as the average number of trees standing at the end of the summer. Figure 7.1 plots the production across various growth rates in a forest with one hundred locations and a lightning probability of 2 percent. Production peaks at a growth rate of 43 percent. The changes in production as growth rates increase is the result of two countervailing forces. The first force is the growth rate itself—as it increases, the more likely that vacant spots are occupied by trees and the greater the potential productivity of the forest. The second force is lightning. As lightning increases, the more likely it is that a tree will be destroyed by fire. Moreover, lightning not only impacts individual trees but contiguous stands of trees. Thus, the density of the trees becomes important. At high growth rates, almost all the trees are contiguous, and all it takes is one strike to bring down the entire forest. This implies a trade-off between these two forces: faster growth means more trees, but more trees imply larger contiguous collections of trees, which promote larger fires.

A two-dimensional (think about trees growing on a checker board) version of the model displays a much more dramatic connection between production and growth. In such a model there is a very dramatic change in production as growth rates are altered.[6] In physics terms, such a dramatic change is known as a phase transition, and it can be shown (via percolation theory) that there is a "critical value" of g that results in the system going from a largely disconnected collection of trees to one in which all the trees are connected together as one.

7.2.3 Homogeneous Adaptation

We now extend the model by considering agents that adapt their rules in a homogeneous manner. We can interpret such homogeneous adaptation as group selection among homogeneous, fixed-rule models. Imagine a collection of forests, each of which has a different growth rate. We impose selection on the system by allowing those forests with higher productivity to survive and those with lower productivity to be killed off and replaced by new forests with random growth rates. Such selection could come about due to natural processes or, in the case of our forest, through the behavior of the U.S. Department of the Interior or some logging company. Over time, such a selection operator will concentrate the forests on those growth rates that lead to high productivity.

In the case of our model, we would expect that selection would push the forests to the critical value of g. In our basic Forest Fire model, production as a function of growth rate is single peaked, achieving a maximum at 43 percent. In such a well-behaved world, all sorts of selection algorithms—from hill climbing via local steps to more elaborate evolutionary methods—should quickly converge to the optimum.

The adaptive solution to the model has an interesting implication: the system adapts to a precipice. Recall that the maximum productivity of this system is associated with a critical value and that such values imply that small deviations can result in a substantial decrease in yield.[7] Thus, adaptation leads the system to a state that is both optimal and fragile.

---

[6] In the one-dimensional model shown in figure 7.1, production tapers off only gradually on either side of the optimal value.

[7] This effect is most dramatic in systems with more than one dimension.

In chapter 8 we discuss, and hopefully provide a bit of needed perspective about, the ideas of "evolution to the edge of chaos" and "self-organized criticality." The result here has a similar flavor to these ideas; adaptation leads the system to a very interesting state that is rich in performance yet rather exposed—nothing ventured, nothing gained. Indeed, the potential for adaptation to drive systems to such a precipitous state is a compelling reason for why adaptation matters in complex systems.

7.2.4 Heterogeneous Adaptation

We extend our model yet again by allowing individual agents to differ in their growth rates. Here we allow agents to select their initial growth rates independently and then adapt them individually. A priori, we should not necessarily expect this model to adapt to the critical value. The system might be too complex, with a tangle of shifting growth rates resulting in an incoherent structure that is impossible for any of the agents to exploit in a productive manner. Alternatively, the system might develop enough structure so that the agents can find productive niches, yet even here the resulting structure might be dramatically different from the uniform growth rates seen earlier.

To construct such a model, we first let each location begin with a randomly assigned growth rate. In each period, we then allow each agent to adjust its growth rate, using the following rule: if a tree on the site would have burned down, then decrease the growth rate; otherwise, increase it. Thus, the agents are in essence conducting a hypothetical experiment at each time period of the form, "If I would have produced a tree, would that have been a good thing or not?" If it would have been a good thing, then increase the growth rate; otherwise, decrease it.

Figure 7.2 shows the mean growth rate of the population over time under heterogeneous adaptation. Initially we see that the mean growth rate declines and falls well below the critical value found in the homogeneous model (43 percent). Eventually this decline reverses itself and the mean growth rate rises and stabilizes at a value (around 59 percent) that exceeds the homogeneous critical growth rate. Throughout these changes in growth rates, forest productivity is steadily rising and eventually settles at a value above the "optimal" value found in the earlier model (production eventually reaches over 65 percent versus our earlier maximum of 58.5 percent).

Clearly, something different is happening in the model with heterogeneous adaptation. To unravel this mystery, we need to investigate the individual growth rates that arise. Table 7.3 shows a representative section of the forest in the later stages of adaptation. Note that the growth rates show an interesting spatial structure: groups of contiguous,

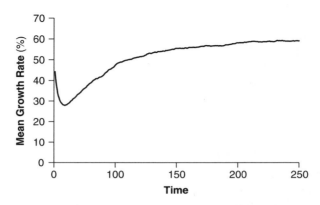

Figure 7.2. Mean growth rate in a Forest Fire model with $f = 0.02$ and heterogeneous adaptation.

TABLE 7.3
Heterogeneous Adaptation in a Forest Fire Model

Location	1	2	3	4	5	6	7	8	9	10	11
Growth Rate	1.00	1.00	0.00	1.00	1.00	0.99	0.00	0.01	0.99	1.00	1.00

high-growth-rate agents are bounded by agents with very low growth rates. Thus, we see that the system has evolved fire walls and, by doing so, has increased overall production above what was possible with a uniform growth rate. This clever solution arose without any kind of central planning and without intent on the part of the individual agents; in Adam Smith's (1776) words, each agent was "led by an invisible hand to promote an end which was no part of his intention." This is a nice example of "emergence" in which an unexpected, higher-level phenomenon arises from lower-level interactions.

Note that the incentive structure we had in place—mediated by the adaptive mechanism—facilitated the development of fire walls. From an economic point of view the resulting solution is an odd one, especially if we assume that having a tree on your site is better than not having one. In particular, the agents that become fire walls provide a very valuable service to those agents that grow trees, yet they are left with nothing. This arises because it is precisely the fire wall agents that are the most vulnerable to fires, as they separate two very fire-prone areas. Given our adaptive system, once the spatial structure begins to form, the fire wall agents will have their growth rates driven to zero, while the neighboring agents will have their growth rates increased. Of course, a "farsighted" fire wall agent could circumvent the adaptive mechanism by recognizing that by setting its growth rate to one, a neighbor would be forced to become a fire wall.

The dynamic system we have explored is fairly typical. Existing sets of behaviors create outcomes that then feed back on those behaviors, creating new outcomes, and so on. Here we see an elaborate dance of behaviors and outcomes, leading the system through a variety of modes and eventually settling down to an equilibrium in which the model resembles a fixed-rule model (but, with heterogeneous rules). In the preceding case, we get behavioral closure that results in an equilibrium. Of course, here we generated the closure through adaptation rather than assuming it outright. This notion of growing it to show it (or, in Epstein's [1999] parlance, generative explanations) may be a nice addition to the desiderata of modeling.

The above exploration begins to hint at why adaptive social systems can fundamentally differ from mere physical systems. Physical systems typically rely on fixed rules, and as we saw in the initial model, such rules may indeed lead to complex behavior. However, adaptation tends to place the system in more interesting regions of that space. Moreover, in adaptive social systems we find that the agents' rules often respond to the phenomena that they generate, creating multiple layers of feedback that result in a diverse set of emergent behaviors, both for the agents and the system at large.

7.2.5 Adding More Intelligence: Internal Models

As a next step, we allow our agents to construct internal models of the system—the fire in the mind. Agents make assumptions about the growth rates of their neighbors and try to set optimal growth rates, given what their neighbors are doing. While the strategy used here is much more sophisticated than the ones used in the previous models, there is no guarantee that it leads to a better solution. It could be that more sophisticated agents result in poorer coordination.

For the moment, consider equilibria where each agent's growth rate is either zero or one. This system has multiple equilibria in the sense that a variety of growth-rate sequences result in agents having no incentive to alter their behavior. Some of these sequences are better than others in terms of system productivity. Moreover, from an individual agent's perspective, equilibria where it produces a tree are better than those in which it serves as a fire wall.

Different internal models will impose different behavioral dynamics on the system. These dynamics result in basins of attraction that will lead any member of that basin to a particular outcome. The issue of the effectiveness of any given learning rule can be recast in terms of how

TABLE 7.4
Optimal Growth Rate Distribution in a Heterogeneous Forest Fire Model

Location	1	2	3	4	5	6	7	8	9	10	11	12	13	14
Growth Rate	1	1	1	1	1	1	0	1	1	1	1	1	1	0

the implied basins of attraction map to the various equilibria. A sophisticated rule could be too greedy and create relatively large basins for the low-yield equilibria. For example, agents, in trying to ensure that they are trees rather than fire walls, may force the system into a perpetual muck in which too many trees are chasing too few fire walls. Lowering the intelligence of these agents might avoid such a dismal fate (as we saw in the case of heterogeneous adaptation).

The idea that imperfection is a productive way to navigate multiple equilibria has been shown in many contexts, such as in simulated annealing. This research indicates that allowing mistakes (especially if they are not too costly or occur early in the search process) helps systems escape less productive outcomes and converge on more productive ones. Less perfection is often more in these types of systems.

7.2.6 Omniscient Closure

Finally, consider a version of the model with omniscient agents. The first question is whether they can achieve the optimal configuration. This requires that the agents can compute the optimal configuration and, once found, get there and stay there. For our 2 percent fire rate, we conjecture that the production-optimizing configuration consists of groupings of six locations with growth rates of one separated by locations with growth rates of zero as shown in table 7.4.

Of course, omniscient agents are not necessarily driven by global optimization. If an agent only obtains value by producing a tree, then that agent would never pick a growth rate of zero, and we will never end up at a production-optimizing equilibria. However, if having a tree burn is worse than not having a tree at all, we can show that the production-optimizing outcome is a possibility with omniscient agents. If the agents all cared a great deal about equity, and side payments were allowed, then the homogeneous outcome at the critical growth rate emerges as the equilibrium.

For the sake of argument, suppose that the configuration in table 7.4 was the omniscient outcome. Given this, how does this outcome compare to the one we found with heterogeneous adapting agents? First, the former model evolved via an observable dynamic process with initial growth-rate fluctuations resulting in an overabundance of fire walls that are only slowly, and imperfectly, eaten away. Second, the adaptive solution has locally stable, but not globally optimal, outcomes with contiguous tree growers in variable-sized groups with from one to six agents, separated by one or even two fire walls, whereas the omniscient solution embodies an elegant symmetry.

The comparative elegance of the omniscient closure model is appealing, but more than likely difficult, if not impossible, to attain in reality. Even if agents could conceive of such a clean solution, the coordination required to jump to it may not be possible. Alas, agents are often left fumbling toward ecstasy, especially in worlds with even a modicum of realism.

7.2.7 Banks

The various transformations of the Forest Fire model demonstrate how we can take a complex systems model and turn it into a complex adaptive social systems model. In the initial model, trees and lightning followed a limited set of fixed rules and did not adapt their behavior. We then introduced adaptation based on selection. By the end of the exercise, trees were embodied with much more cognition, and we might eventually contemplate a system whereby the trees begin to reason "I'd better grow more slowly, as the current El Niño portends an increase in lightning next summer."

As we add adaptation to the Forest Fire model, we can begin to extend its applications to more social phenomena, such as bank failures, the spread of human diseases and computer viruses, and the spread of information. Consider the case of bank failures. Such failures can cause widespread economic disaster, as was seen in 1997 when a cascade of financial-system problems devastated a number of Southeast Asian economies or when in 1998 the Federal Reserve organized a $3.5 billion dollar bailout of Long-Term Capital Management to prevent a failure cascade that could have reverberated across the world's financial markets.

In a Bank Failure model, each tree (the "fuel" for the fires) is analogous to a risky loan extended by a bank, and the topology of the forest represents connections between banks mediated by correlated risks among their loans.[8] To complete the analogy, lightning strikes represent loan defaults, and spreading fires capture the cascades of resulting bank

[8]Actual interbank relationships create a much more elaborate topology of interaction than what we consider here, but accounting for this requires only modest changes in the model.

failures as loan calls between banks go unanswered. Regeneration is the tendency for banks to make new loans.

Model transformations can sometimes be quite subtle. For example, consider a model of agents who can either stay in the country or move to the big city. If they move to the city, they meet up with any kinfolk and remain there unless they, or anyone they know, gets mugged, in which case they move back to the country. You can picture the dynamics of such a model, with agents moving back and forth between the city and country, muggings, and so on. While on the face of it such a model would appear to require a completely different apparatus than that developed here, in fact it is isomorphic to our Forest Fire model, with the growing of a tree being equivalent to an agent moving to the city, fires becoming muggings, and neighboring trees giving kin connections.

7.3 Eight Folding into One

Notwithstanding the tidy taxonomy presented in section 7.1 (or any taxonomy for that matter) or the various explorations of the Forest Fire model in section 7.2 (or any other model), there is a fundamental problem in classifying and characterizing any agent-based computational model—namely, at the deepest level all such models have a similar formulation, and it is only the observer's frame of reference and understanding that differentiates them in any real sense. Alas, at some levels our noble path can easily collapse into nothingness. The reason for this is that all of the models we focus on are ultimately embodied as a fixed, nonstochastic algorithm.[9] As such, at the deepest level there is no adaptation, heterogeneity, space, and so forth inherent in these models, just a fixed rule going through its paces being interpreted in various ways by an observer.

There is a universality among computational systems in which, once a certain threshold is passed, each such system is capable of performing the other's computations.[10] Thus, balls colliding with one another on a billiard table, *suitably arranged and interpreted*, can perform the same computations as a supercomputer or any possible agent-based model. This implies that it is possible for a dumb system (colliding billiard balls) to emulate a smart one (sophisticated agents interacting in a social system).

---

[9] Even randomness in these models is induced via a pseudorandom number generator that is fixed modulo a seed.

[10] Wolfram's (2002) principle of computational equivalence uses this idea to claim that things can only get so complex—once you make it over the threshold, you are only as good as the next guy.

Table 7.5
A Simple, Nearest-Neighbor Cellular Automaton

Input State	Output Choice
000	0
001	0
010	1
011	1
100	0
101	0
110	1
111	1

Of course, this universality requires that we know how to set up the initial conditions and how to interpret the states. For most problems, it is likely to be extremely difficult to derive suitable initial conditions that will carry out the computation. Even in those cases where we know the initial conditions and all of the rules that govern the dynamics, we may still not be able to predict or understand what will happen to the system. As we move away from being omniscient observers, the system becomes increasingly complex and adaptive.

To illustrate some of these ideas, we consider the behavior of a simple cellular automaton. In such a model, two-state agents (for simplicity, agents can either be in state 0 or 1) are arrayed around a circle. At each time step, agents look at their own state and the states of their nearest neighbors and, based on this information, determine their next state. We can capture this behavior in the form of a rule table like that shown in table 7.5. A rule table is a mapping from each possible input state to an output state. Here, the input state is ordered by the state of the left neighbor, the agent, and the right neighbor, respectively. The rule in table 7.5 implies that the agent will ignore the actions of its neighbors and just repeat its own state at each time step. If all the agents use this rule, the system locks into a fixed, repeating pattern that is directly tied to the initial configuration of the system.

From the perspective of an omniscient observer, this system is rather dull. A group of homogeneous agents, using fixed rules, immediately lock into a deterministic pattern, which is directly tied to the initial conditions. Of course, even this "boring" behavior can be imputed with meaning; for example, this system could be interpreted as a decentralized storage device that remembers whatever initial state is given to it.

From the perspective of agents within the system, the story is a bit more complicated. Even though each agent's behavior is based on an identical, fixed rule, there are really two "types" of behavior linked to each agent's initial state. Agents that begin in state 0, follow a behavioral program that is determined by the first, second, fifth, and sixth rows of the rule table, while those in state 1 use the remaining rows. Thus, the internal state of each agent embraces a very different behavioral program, and from "one" rule table we see two behaviors. If, by chance, the internal state changed, then it would appear to a neighboring observer that somehow the agent has changed its behavior.

Thus, depending on the information available to an observer, we can interpret the identical system in very different ways. From the perspective of an omniscient observer, who has access to the agents' states, rules, and initial conditions, the system is composed of a collection of homogeneous, fixed-rule agents. From the perspective of an agent within the system, unable to observe the internal states of its neighbors,[11] it looks like the system is composed of heterogeneous agents. Thus, the frame from which agents are viewed—in particular, the knowledge of the overall states, rules, and initial conditions—can radically alter our classification of a system.

To push this latter point just a little bit further, consider the following. It is known that other rules like that in table 7.5 can have more elaborate behavior. Thus, by changing three output choices in the table (the bits for the input states 001, 101, and 111), we get a new rule that Cook and Wolfram showed to be a universal computer (Wolfram, 2002). Thus, assuming we can find the appropriate initial condition and interpret the output, we could emulate any agent-based model with a suitably sized automaton using this latter rule. Therefore, all agent-based models could be formulated as a homogeneous, fixed-rule, nearest-neighbor automaton.

The oneness that results from all of these systems relying on fixed, deterministic algorithms produces an interesting scientific conundrum. At the deepest levels, the systems we are interested in could not be simpler; yet, given our limited observational powers, they still can be filled with wonder and puzzles. Moreover, all such models are connected to one another. If we can predict and understand how a universal computer behaves, then we can predict and understand all agent-based models.

Unfortunately, we know from Turing's (1937) insights into the Halting problem that *general* propositions about universal computers are undecidable. That is, it is not possible to make general statements about, say, whether a model based on computational agents will reach an equilibrium or not. Of course, such general statements may be asking for too much, and we may be satisfied with knowing the

---

[11] We recognize that here internal states are directly observable by neighbors, but for the sake of rhetorical brevity, we will ignore such a complication.

answer to such questions for a particular set of systems (which may be possible); nonetheless, the halting problem does point to some inherent limits to theory in this domain.

7.4 CONCLUSION

Our analysis of the Forest Fire model has enabled us to see how different levels of adaptation can impact behavior. We started with a standard fixed-rule model and then allowed for successive degrees of adaptation. During this exercise we saw the emergence of fire walls, which showed how a collective intelligence can arise without intent on the part of the agents. The adaptive system generated an efficiency-increasing, though baroque, solution to the problem at hand. Finally, when we allow omniscient closure, a starkly symmetric outcome results, though the actual ability of a real system to generate this outcome is suspect.

There is something far more intriguing inherent in the dynamics that arise in between the fixed-rule and cognitive closure models. This interesting "in between" where agents fumble for a solution and out of this process something both clever and messy emerges is a recurrent theme in subsequent models.

We also saw how transforming the labels of the Forest Fire model can allow us to explore something like bank failures. In the example of country-city migration, we were able to use the same model to account for a phenomenon that seem to go well beyond simple relabelings. Such transformation is important for successful modeling, as it allows efforts in one domain to be conserved and reused, while also identifying some deeper connections across seemingly disparate systems. The ability to transform one model into another, and the theorems that imply that all of these types of systems are intimately connected, suggest that a deep understanding of a few agent-based models might yield a much broader understanding of many other social systems.

At the deepest level there is a fundamental sameness about all of these models, as they all are based on a discrete, dynamic system cycling through various states. Of course, if we knew everything about how this latter system behaved, then our task would be considerably easier, as we would just have to map the primitives of the social systems we wish to understand to those of the universal system that we understood. Unfortunately, we do not yet have this deeper understanding of this universal system; moreover, we know that there are some inherent limits to theorizing in such a system. Nonetheless, the promise of uncovering deep connections among apparently disparate complex adaptive social systems is an important one.

CHAPTER 8

Complex Adaptive Social Systems
in One Dimension

Be patient, for the world is broad and wide.
—*Edwin A. Abbott, Flatland*

WE BEGIN WITH A SET of very simple models designed to illuminate some basic issues inherent in complex adaptive social systems. In Abbott's *Flatland*, geometric figures confined to living in a two-dimensional world gain insight into the third-dimension when a sphere slowly passes through their plane. The sphere begins as a point, grows into ever larger circles, eventually reverses its course and returns to a point, and disappears. After seeing this amazing sequence of activity, the figures confined to Flatland begin to glimpse the third dimension. Here we explore some simple models with a similar motivation to Abbott's sphere, namely, to provide some useful glimpses into the behavior of complex adaptive social systems.

Modeling any system is often an exploratory process that requires both induction and deduction. You begin by making a simple set of assumptions and see where they lead. From this experience you attempt to create better models or deduce more exact results. The discussion in this chapter follows this approach. We begin with a rather stark notion of a social system of interacting agents and then attempt to direct the analysis down productive paths. Our goal is not only to illustrate how such models can be developed and analyzed, but also to create a series of easily digestible models that embody many of the key concepts and insights that have been developed in complex adaptive social systems over the past decade. These devilishly simple models are not some random stroll through the set of possibilities but contain a significant malice of forethought.

The agents in the models presented populate a one-dimensional circular world—for concreteness, consider a world in which agents live atop a large atoll. Around this atoll we have N sites that can be occupied by the agents. For ease of exposition, we consecutively number these sites $1, 2, \ldots, N$ starting from an arbitrary point, and thus the Nth site completes the circle and abuts site 1. Given this world, we can impose a natural constraint on agent interactions, namely, that agents interact within neighborhoods of contiguous sites. Thus, an agent at site 1 has a "right-hand" neighbor at site 2 and a "left-hand" neighbor at site N.[1]

In the initial models, agents must take one of two possible actions (designated by 0 and 1). Each agent chooses its action using a fixed behavioral rule. This behavioral rule depends only upon the most recently observed actions taken by the agent and its designated neighbors. While this construction is obviously very stark, it is sufficient to demonstrate a number of the core features that arise in complex adaptive social systems and computational modeling.

A crucial difference between models of complex social and physical systems is in our assumptions about appropriate behavioral rules. Quantum effects aside, one hydrogen atom acts just like another hydrogen atom relying on a set of fixed, external physical properties and forces. Social agents, on the other hand, often alter their behavior in response to, and in anticipation of, the actions of others.[2] The atoms on the bumpers of two cars about to collide do not alter their behavior; the drivers of the two cars typically do, albeit a bit late. As a result, social systems have an additional layer of complication over physical ones, and we must make sure that the behavioral rules deployed by our agents make sense in this broader context. We initially explore systems with simple fixed rules to gain some basic insights and intuitions. Eventually, we introduce more complex rules that can "change" their behavior over time.

One can often think of complex adaptive systems as having microlevel agents (in the case of our atoll, "Micronesians") interacting to create the global properties of the system. These global properties then feedback into the microlevel interactions in various ways. Such feedbacks occur in both physical systems, like earthquakes, and social ones, like stock market crashes. What differentiates physical systems from social ones is that agents in social systems often alter their behavior in response to anticipated outcomes. Rocks on the boundaries of tectonic plates just let earthquakes happen; people attempt to prevent stock market crashes.

8.1 Cellular Automata

The first model we explore is one in which each agent's behavior is driven by the same generic rule. Consider an atoll of size 20, where each site is occupied by an agent that has two possible actions {0, 1}. We assume that

---

[1] We can extend this idea to larger neighborhoods as well. An agent at site 1 has immediate neighbors at sites 2 and N, neighbors two steps away at sites 3 and $N - 1$, etc.

[2] As we discussed previously, even "adaptive" rules are fixed at the deepest levels.

TABLE 8.1
A Simple Behavioral Rule

Situation	Left Neighbor	Self	Right Neighbor	Rule 22
0	0	0	0	0
1	0	0	1	1
2	0	1	0	1
3	0	1	1	0
4	1	0	0	1
5	1	0	1	0
6	1	1	0	0
7	1	1	1	0

each agent's behavior is controlled by the *identical* rule, and that this rule uses the most recent action of the agent in question and its two nearest neighbors to determine the next action. Given that actions are binary, a fully specified rule will need to map the 8 (2^3) possible combinations of actions that the agent and its two neighbors can take, into the agent's next action. Because the rule must designate a binary action (either 0 or 1) for each of the eight situations, there are 256 (2^8) possible rules that could direct an agent's behavior.

Table 8.1 shows the rule table for Rule 22.[3] A rule table is a mapping from each possible input state to an output state. The first line of this rule table (situation 0) specifies that if an agent and its two neighbors all took action 0 last time, then the agent will want to take action 0 next period. The next line (situation 1) indicates that if the agent and its left neighbor took action 0 and the right neighbor took action 1, then the agent will want to take action 1, and so forth. In table 8.2 we show the dynamics of this simple rule. The sites of the atoll are numbered from 1 to 20 moving left to right (recall that the left- and right-hand edges of the table are connected to one another, so a more accurate representation would entail forming the table into a cylinder by rolling the outer edge of the page in toward the binding). At time 0, we randomly pick an action for each agent. At each subsequent iteration the agents simultaneously choose their next actions based on the rule in table 8.1 and the actions observed in the previous time step.

This simple rule results in some interesting systemwide behavior. As can be seen in table 8.2, coherent macrostructures in the form

[3]A standard way of referring to such rules is by using the integer equivalent of the bits that define the rule table. Thus, in table 8.1 the defining bits of the rule table are 00010110, which can be interpreted as the integer value 22.

Table 8.2
Dynamics of Rule 22

Time Step	Actions
0	11100100000111011100
1	00011110001000000011
2	10100001011100000100
3	10110011000010001111
4	00001001001110100000
5	00010011111000011000
6	00111100000100100100
7	01000010001111111110
8	11100111010000000001
9	00011000011000000010
10	00100100100100000111
11	11111111111110001000
12	00000000000001011101
13	10000000000011000001
14	01000000000100100010
15	11100000001111110111

of downward facing triangles composed of 0s emerge throughout the diagram. The scale of these triangles goes well beyond the scale of the behavioral rules. Thus, even though individual behavior is based on the actions observed at three sites, coherent triangular structures emerge that encompass far more sites (for example, at time 12 a triangle forms across thirteen of the twenty sites). Like Adam Smith's invisible hand, it is as if the actions of the agents are being coordinated to create patterns that are no part of any agent's intention.

While there is some coherence in the outcome, there is also perpetual novelty. Thus, while the system has a "theme" of a recurring series of downward facing triangles, their sizes and locations seem to vary across space and time in such a way that we never seem to see the exact pattern twice. This latter point needs to be qualified. Our system has only a finite number of possible states—with an atoll of size 20, there are 2^{20} (a little over a million) possible unique configurations of the agents' actions. Because the rules are deterministic, any particular configuration is always followed by the same subsequent configuration. Therefore, if we run the system long enough (at most 2^{20} time steps), it is guaranteed to hit the same configuration twice, and once this happens it will begin a cycle that follows the same path as it did when it first hit the configuration. All finite, deterministic systems are guaranteed to cycle, though the lengths of these cycles can be relatively long.

The preceding rule demonstrates how simple, local interactions among agents can result in interesting aggregate behavior. The rule is just one of 256 possible rules, and an obvious question is whether the behavior we see in this rule is in some sense generic. The answer is no. For example, a rule table where each possible situation results in a 1 will immediately lock the system into an equilibrium where all agents do action 1 after the first time step. Alternatively, a rule that always has an agent doing the opposite of what it did last period (that is, having 0s in situations 2, 3, 6, and 7, and 1s elsewhere) will cause the system to alternate back and forth with each time step.

Wolfram (1984b, 2002) has systematically analyzed the 256 possible rules and divided their behavior into four classes. Class 1 rules quickly evolve to a unique, homogeneous state with identical actions across the agents (as in the "all 1s" rule). Class 2 rules result in separated groups of simple stable or periodic structures (as in the "do the opposite" rule). Class 3 rules imply chaotic patterns (the rule in table 8.1 is a member of this class). Class 4 rules produce complex structures with long transients (thus, coherent patterns arise that can persist across space and time for extended periods) that are hypothesized to be capable of universal computation—that is, able to compute anything that can in principle be computed (Rule 110 meets this criterion).[4] One way to quantify the above classes is to measure how a random alteration of an action alters the behavior of the system in subsequent time periods. In Classes 1 and 2, such impacts are minimal, while in Classes 3 and 4 such disturbances can propagate across vast distances.

Wolfram's classification scheme allows us to abstract away particular details of the rules and still make good predictions about aggregate behavior. There are, however, problems with his classification scheme. In particular, the outcome of any given rule depends on both its structure and the initial conditions. It is possible for the behavior of a single rule to fall into two different classes. For example, in table 8.3 we show the behavior, under two different starting conditions, of a rule that copies whatever the left neighbor did last period. The initial conditions in World 1 lead to Class 1 behavior, whereas those in World 2 place the rule within Class 2. Rules can also start out in one class (by, say, displaying a very long transient) and then fall into a different class (by, perhaps, converging to a low-period limit cycle). Notwithstanding these difficulties, Wolfram's attempts at classification represent an important step in creating more general theories of complex systems.

---

[4]Analogs to continuous dynamic systems exist for the first three classes. Respectively, these are limit points, limit cycles, and chaotic attractors.

TABLE 8.3
Copy-Left Rule under Two Different Initial Conditions

Time Step	World 1	World 2
0	1111111111	1000000000
1	1111111111	0100000000
2	1111111111	0010000000
3	1111111111	0001000000
4	1111111111	0000100000
5	1111111111	0000010000
6	1111111111	0000001000
7	1111111111	0000000100
8	1111111111	0000000010
9	1111111111	0000000001
10	1111111111	1000000000
11	1111111111	0100000000

8.2 Social Cellular Automata

So far, we have demonstrated how simple systems of interacting agents modeled by cellular automata can result in interesting behavior. To convert these automata systems into models of social systems, we need some additional qualifications.

The first qualification is that we are willing to accept the notion that all agents employ a common, fixed rule. Many models of social systems embody such behavior by assuming a single, "representative" agent. Even when multiple rules across agents are possible, homogeneity can still arise through a variety of processes. For example, if all agents optimize the same problem in the same way by, say, adopting a Nash equilibrium strategy, then their behavioral rules may coincide. Behavior can also be coordinated by other social forces that may be driven by optimization. For example, behavior (perhaps due to optimization) from one circumstance, such as driving on the right side of the road, may be the basis for behavioral norms in other circumstances, such as veering to the right when confronting another pedestrian on the sidewalk.[5]

[5] While in this case such meme-based hitchhiking serves a useful purpose (preventing collisions in a variety of circumstances), there could be circumstances where it is harmful. For example, norms of sharing may improve an agent's performance in some contexts like dealing with family members, yet be detrimental in other ones such as one-shot anonymous social dilemmas. One focus of behavioral decision research is in uncovering key behavioral heuristics—often by finding circumstances under which they result in wildly maladaptive behavior.

Even when all agents begin by using the same rule, mechanisms are still needed to prevent adaptive agents from deviating away from this rule. In stable environments, where feedback is consistent with good performance and expectations, adaptive agents are not likely to want to change their behavior. In less stable environments, the ability of an agent to alter its behavior productively depends on a number of factors. First, the agent must be able to obtain sufficient information from the observable patterns of behavior and outcomes to formulate more productive plans, and in some environments such patterns may be very difficult to divine. Second, the agent must know about other behavioral possibilities. In later models we explore the behavior of systems with heterogeneous and adaptive rules, but for the moment we assume that conditions are such that agent behavior is described by a static, homogeneous rule.

Another problematic assumption in the preceding model for social situations is that agents myopically apply their behavioral rules to the actions observed last period. Thus, either agents are incapable of remembering and processing more elaborate histories or the actions of the last period are a sufficient statistic of the past (in essence, they incorporate all the information needed for predicting the future). A final qualification is that we assume that the timing of behavior in these models, namely, that all agents update their actions simultaneously, is sufficiently close to real systems (or of little consequence to the outcome of the model). We explore some specific issues surrounding timing later on in the analysis.

These qualifications are certainly not trivial, though they are well within the usual bounds of many social science models. Ultimately, our willingness to entertain them is tied to the value of the subsequent models.

8.2.1 Socially Acceptable Rules

A key area for refining the preceding class of models is thinking about socially acceptable behavioral rules. Indeed, by applying some simple constraints on social behavior we can dramatically reduce the set of admissible rules. There are many ways we could constrain social behavior, and we begin by assuming that social agents have some degree of rationality and are goal oriented. The first constraint we impose, *observational symmetry*, is that symmetric observations of the left- and right-hand neighbors lead to the same action, that is, that agents do not differentiate between their neighbors. As shown in table 8.4 this constraint forces the actions taken in situations 1 and 4 to be identical, as

TABLE 8.4
Social Symmetries in Rule Tables, Where $a_i \in \{0, 1\}$

Situation	Left	Self	Right	Observational Symmetry	Outcome Symmetry	Both
0	0	0	0	a_0	a_0	a_0
1	0	0	1	a_1	a_1	a_1
2	0	1	0	a_2	a_0	a_0
3	0	1	1	a_3	a_1	a_1
4	1	0	0	a_1	a_2	a_1
5	1	0	1	a_4	a_3	a_2
6	1	1	0	a_3	a_2	a_1
7	1	1	1	a_5	a_3	a_2

well as those in situations 3 and 6. By imposing observational symmetry, we go from 256 to 64 possible rules.

Another possible constraint we could impose is that of *outcome symmetry*. Assume that each agent shares a strict ordering of the potential configurations of the actions taken by itself and its two neighbors, and that each agent myopically believes that its neighbors will not alter their actions next period. Under observational symmetry, an agent should not take an action that would lead to different predicted outcomes when identical ones are possible. For example, whatever action, a, the agent takes in situation 000 should be identical to the action it takes in 010, as the agent will myopically predict that it will find itself in $0a0$ next time. The outcome symmetries implied by this constraint are shown in table 8.4. Note that outcome symmetry is a very strong assumption, as it requires a strict ordering across the absolute outcomes, and later we will explore some plausible social models that violate this constraint.

By imposing both observational and outcome symmetry, we dramatically reduce the space of acceptable social rules. The last column in table 8.4 shows the acceptable permutations under both types of symmetry. Note that only three binary values are needed to complete the associated rule table, and thus only eight rules are possible. In fact, given that we can relabel actions by swapping the 0s for 1s and vice versa, we only need to examine four rules. In table 8.5 we show these four prototypical social rules and their canonical names.

These four rules imply very different system behavior. Rule 0 is the least interesting from a social perspective, as the agents always want to take action 0 regardless of what their neighbors are doing. Under this rule the system locks into all agents taking action 0 after the first time step.

TABLE 8.5
Symmetry Constrained Social Rules

Situation	Left	Self	Right	Rule 0	Rule 5	Rule 90	Rule 160
0	0	0	0	0	1	0	0
1	0	0	1	0	0	1	0
2	0	1	0	0	1	0	0
3	0	1	1	0	0	1	0
4	1	0	0	0	0	1	0
5	1	0	1	0	0	0	1
6	1	1	0	0	0	1	0
7	1	1	1	0	0	0	1

Rule 160 is a bit more complicated as each agent wants to take action 0 unless both of its neighbors are doing action 1, in which case, it will join them. Plausible social situations for this behavior might include agents choosing a technology (action 0 or 1) in the presence of network externalities (where technology 0 has an economic edge *ceteris paribus*) or agents forming an agreement and either abiding by it (action 1) or violating it (action 0). There are three possible equilibria under Rule 160. If all of the agents start by taking action 1, then they will continue this action throughout all subsequent iterations, and the system will lock into all of the agents doing 1. A second possibility is that the system locks into a configuration with actions alternating 0101... across the sites (which, of course, requires an even number of sites on the lattice). In this case, agents will alternate their actions during each iteration, and the system will cycle back and forth with period 2. This alternating configuration requires a fairly delicate initial condition consisting of one or more noncontiguous 0s with distances between them that are even numbers.[6] Finally, if it is ever the case that two or more contiguous agents take action 0, then during each subsequent iteration the neighbors to this contiguous group will also take action 0 (and the agents within the bunch will stay with 0), and 0s will slowly propagate throughout the lattice, eventually locking the system into all of the agents taking action 0. Note that this last result requires initial conditions with either one or more groups of contiguous 0s or isolated 0s spaced an odd number of sites apart. Notwithstanding the three possible equilibria, under random

---

[6]To see this, consider a single 0 surrounded by 1s. At each subsequent time step, an alternating pattern of 01s will propagate out by one. With multiple, noncontiguous 0s, then these patterns will join up as long as the initial 0s are spaced an even number of sites apart.

starting conditions or, say, in the presence of noise, the last equilibrium of all 0s is the most likely outcome.

Under Rule 5 agents want to take action 0, unless they can be the only one in their neighborhood taking action 1. Social situations captured by such incentives might include getting a painful body piercing or investing in easily appropriated research and development (in each case, the value of taking the action increases with its uniqueness). Systems controlled by Rule 5 are characterized by stable (across time steps) bands of 010, with the gaps between these bands alternating between all 1s and all 0s each time step. The gaps can be of various sizes, so multiple equilibria are quite likely in these systems. Thus, the prediction here is that there will be some stable isolates, where, say, an agent with a piercing is surrounded by two neighbors without, next to groups of agents that experience "fads" that alternate between everyone having a piercing and no one having one. The dynamics of these fads are driven by agent myopia—we might expect that slightly more adaptive agents would be able to recognize and respond to the small-period cycles that are observed in the system, though the impact of such a response is not clear a priori.

Rule 90 often results in exotic behavior. It is a Wolfram Class 3 rule[7] and thus can exhibit chaotic patterns that are similar in structure to those shown in table 8.2 (but with even more complicated "triangles"). Behaviorally, this rule models an agent that wants to be in a neighborhood either with exactly two 1s or none. Such behavior is consistent with the rules used in Conway's Game of Life (a model based on two-dimensional cellular automata), in which an agent becomes alive (action 1) if there is a parent and sufficient space, and dies (action 0) if the world is either too crowded or too lonely. A more economically oriented example could be our atoll inhabitants needing to have access to at least two boats (action 1) to go fishing with their neighbors.

The preceding analysis indicates that, if we are willing to consider a sparse and highly constrained model of interactive social agents, we are likely to observe only four types of generic system behavior. Two of these systems result in stable and highly predictable equilibria that are quickly attained by the system. The third system has the possibility of multiple equilibria, but again the characteristics of these equilibria are generic and we are likely to observe an easily recognizable two-period oscillating pattern where bands of agents "overreact" to past information. The final system is one that is rich in behavior and pattern, though more exacting predictions are difficult to make.

By putting some sensible, albeit extreme, simplifications on the problem, we were able to create a very small set of relatively easily analyzed,

---

[7]Rule 5 is Class 2, and Rules 0 and 160 are both Class 1.

TABLE 8.6
A Nearest-Neighbor Majority Rule

Situation	Left	Self	Right	Majority Rule
0	0	0	0	0
1	0	0	1	0
2	0	1	0	0
3	0	1	1	1
4	1	0	0	0
5	1	0	1	1
6	1	1	0	1
7	1	1	1	1

yet interesting, models of interacting social agents. Obviously, we do not wish to claim that the behavior of all complex adaptive social systems can be subsumed by one of the four generic types of outcomes we uncovered. Rather, the work is meant to be illustrative of a style of modeling in this area and its potential for providing new insights.

In the next sections we develop some alternative models of complex adaptive social systems. In these models we loosen some of the constraints in various ways, so that we can gain insight into the importance of some of our assumptions as well as investigate new elements of social behavior.

8.3 MAJORITY RULES

In the Majority Rule model, we assume that agents attempt to take actions that are consistent with the majority of their neighbors.[8] Thus, each agent will look at all neighbors within a distance of k sites and alter its action if it is in the minority. In table 8.6 we write down the rule table for this behavior when $k = 1$. Notice that this rule is not a member of the "socially acceptable" class developed earlier, because it violates outcome symmetry in those situations where neighbors are taking opposite actions from one another (here, agents are willing to stay with whatever their previous action was, perhaps due to switching costs, in those cases where they are pivotal).[9] If we impose a preference on the agents by, say, making them always wanting to be part of a majority and, if possible, having that be a majority of 0s, then their behavior is driven by Rule 160 discussed in the previous section.

---

[8]This type of model is often referred to as a Voter model in the literature, but we will use the more exacting term Majority Rule model.

[9]An agent is pivotal when its choice determines the majority.

TABLE 8.7
Majority Rule ($k = 3$) with Synchronous Updating

Time Step	Actions
0	00111011100111001000
1	00011111111101100000
2	00011111111111000000
3	00011111111111000000

In table 8.7 we show some typical behavior for the model with $k = 3$, that is, when agents look to their three left and three right neighbors for guidance in determining their next action. The dynamics shown in the table are very typical—the system quickly settles down to a world with stable blocks of 1s and 0s. As long as there are contiguous blocks of the same action that are at least $k + 1$ sites in length, the system will be in equilibrium, since under this condition every agent has at least k like-minded neighbors and thus will not want to alter its action. Of course, the actual number, location, and ultimate size ($\geq k+1$) of each block depends on the initial conditions, though, as we show later, there are some useful statistical regularities to these features. It is possible, although not very likely, for a periodic equilibrium to occur in this system. If we have an even number of sites and the initial configuration alternates between 0 and 1, then each agent's action will alternate from one time period to the next.

One implicit assumption in the models so far has been that all agents update their actions simultaneously. As previously discussed, sometimes decentralized systems of agents might attain high degrees of coordination through natural or artificial cues such as sunrise (when many people make clothing choices) or legal mandates like election days. Notwithstanding such examples, it is easy to imagine situations where agents take asynchronous actions. Once we allow for asynchronous actions, we must define an activation order for the agents. Agents can activate randomly or according to some order driven by spatial location, endogenous agent characteristics such as age, or even incentives for taking action.

In table 8.8 we show some sample equilibria under various updating rules. Each of these rules began with the same initial condition. Under asynch-location, the agents are updated moving in order from site 1 to site 20. In asynch-incentive, agents with the fewest number of like-minded neighbors update first. Finally, in asynch-random, agents are updated randomly with replacement. Under all of the updating rules, the system reaches an equilibrium (characterized by the previously

TABLE 8.8
Majority Rule ($k = 3$) Equilibria under Various Updating Options

Updating	Equilibrium
Initial condition	00111011100111001000
Synchronous	00011111111111000000
Asynch-location	00000000000000000000
Asynch-incentive	00111111111111000000
Asynch-random	00011111100000000000

TABLE 8.9
Average Number of Equilibrium Blocks

| Updating Rule | k | | | | | | | |
---	1	2	3	4	5	6	7	8
Synchronous	834	478	342	266	215	178	152	133
Asynch-location	833	358	162	80	41	22	8	4
Asynch-incentive	922	488	307	214	161	143	101	80
Asynch-random	916	492	326	243	196	162	140	123

Note: Equilibrium blocks form when stable configurations of contiguous, identical behavior, result from the dynamics. The experiment had $N = 5,000$ and 50 trials.

mentioned stability condition of contiguous, homogeneous actions of size greater than $k + 1$) relatively quickly.

Table 8.8 suggests that different updating rules may result in very different equilibrium outcomes.[10] The specific outcomes depend on both the initial conditions and random elements, so it is hard to make any useful generalizations without more observations. In table 8.9 we show the average number of blocks that form in equilibrium across fifty trials on a 5,000-site lattice under the different updating conditions.

The statistical results provide fodder for a variety of theoretical explorations. For example, notice that under location-based updating, each increment of k results in roughly a halving of the number of equilibrium blocks. A simple explanation for this phenomenon relies on the following insight: if we are moving down the lattice from left to right with a homogeneous block behind us, we will start a new block if and only if the next site we encounter and *all* k of its neighbors to the right are taking the alternate action. The probability of this happening with

[10]This result is consistent with the observations of Huberman and Glance (1993) and Page (1997).

randomly generated sites is $(1/2)^{k+1}$, and thus incrementing k by one decreases this probability by one-half.

In all of our numerical experiments with this system, we noticed that asynchronous updating led to a noncyclic equilibrium. We know that with synchronous updating such equilibria are likely, though it is also possible to sustain a two-period cycle. Are noncyclic equilibria a fundamental feature of this system under asynchronous updating?

Claim 8.3.1 *Any N-site, k-Majority Rule model with asynchronous updating attains a fixed-point equilibrium in finite time.*

Proof: To prove this claim, let n_i^t give the number of neighbors of site i at the beginning of time t that are taking a different action. We can construct a (Lyapunov) function, F, that is given by $\sum_i n_i^t$, that is, the sum across all agents of the number of neighbors that disagree with the agent's current action. Note that this sum is finite and bounded below by zero. Suppose that at time t, agent i is given a chance to update. There are $2k$ neighbors of site i, and the agent will alter its action if it is in the minority, that is, if $n_i^t > k$. If the agent is not in the minority, no changes occur on the lattice and the value of F remains unchanged. If, instead, the agent is in the minority, the agent will alter its action to that of the majority. In this case, F will decrease due to two effects. First, the change in the agent's action will alter the value of n_j^{t+1} for each agent in the neighborhood. Those agents in the majority will each have their differences from neighboring agents decrease by one, while those in the minority will have their differences increase by one, resulting in the sum of the differences across all of these sites going down by at least two $(n_i^t - (2k - n_i^t) \geq 2)$. Second, the agent that altered its action will have its difference decrease by at least two, as prior to switching actions its difference was n_i^t and after the switch it is $2k - n_i^t$. Therefore, F must fall by at least four whenever an agent switches. Because F is both finite and bounded below by zero, agents cannot switch indefinitely, and thus the claim must hold.

Mathematical and computational approaches provide complementary insights into this problem. The original computations produced some very useful observations about how the system behaves, including insights into the likelihood of equilibria, their form, and the speed of convergence. More extensive statistical analyses, like those in table 8.9, suggested new mathematical directions as well as defined the potential for empirical work on these types of systems (for example, knowing that a system has, say, 160 blocks in equilibrium may not allow us to recover its underlying structure, since such an observation is consistent with k's ranging from three to seven depending on the updating mechanism).

Mathematical results solidify some of our computational observations. Results like claim 8.3.1 nicely formalize some of our computational intuitions, though they cannot escape the bounds of their assumptions (here, for example, the claim fails under synchronous updating). Also, we have been unable to characterize mathematically some of the most interesting aspects of the system, such as the block formation dynamics (as seen in table 8.9) or time to convergence. That is not to say that such things cannot be so characterized, as they may well succumb to mathematical analysis (though the computational results may suffice for many purposes). Of course, caution is always needed in interpreting both computational and mathematical results. Inductive observations about computational systems and mathematical deductions can be misleading if we do not carefully consider their underpinnings.

8.3.1 The Zen of Mistakes in Majority Rule

The rules we have considered up to this point have been deterministic. Of course, rules can also have random elements. Randomness is not solely within the social domain; for example, neurons fire probabilistically based upon chemical levels. In social systems, randomness can capture features like mistakes, experimentation, or the tendency to bias choices imperfectly.

Note that randomness does not necessarily imply "random" behavior. Randomness is often a source of order in complex systems. For example, consider an agent, randomly placed on a rugged landscape, attempting to climb uphill in a dense fog. If the agent strictly follows an uphill path, it will quickly get trapped on whatever local peak it started on. Now, introduce some randomness into the system whereby the agent has some chance of taking downhill steps during its fog-bound excursion. Such noise will start to "smooth" out the landscape by "filling in" the minor valleys that separate local peaks, allowing the agent to transit the landscape's minor undulations. Thus, by introducing some noise the agent will gravitate to the highest peak in the landscape (a very orderly outcome) rather than becoming trapped in its initial neighborhood (a very disorderly outcome). The intuitions in this example form the basis for interesting optimization ideas like simulated annealing.

When we add noise to the Majority Rule model, we get different results. Suppose that agents who lie between agents taking different actions make mistakes and that those surrounded by similar agents do not. In essence, this implies that little mistakes can happen but big ones do not. Such an assumption is similar to the concept of Proper

TABLE 8.10
Majority Rule ($k = 3$) with Mistakes

Updating	Equilibrium
Initial condition	00111011100111001000
Convergence	00111111111111000000
After some mistakes	11111111111110000000
An equilibrium	11111111111111111111

equilibrium in game theory, in which the likelihood of a mistake is proportional to its costs.

Suppose that we run our Majority Rule model without noise using synchronous updating until it converges, as shown in the second line of table 8.10. This configuration is not stable under noise, as the four agents spanning the two boundaries will make mistakes, in the sense that with some small probability each may switch its action. These mistakes will start to eat away at the edges of the long strings of contiguous values. Eventually, the system will reach a state of either all ones or all zeros, at which point mistakes will end, and an equilibrium will ensue.

With errors we find that majority rule leads to either all ones or all zeros. Moreover, if we were to average over all possible initial strings, then those with an initial majority of, say, ones would be more likely to converge to the equilibrium of all ones. This occurs because longer sequences of the majority symbols are less likely to get wiped out due to mistakes than shorter ones. This simple example shows how mistakes can be beneficial. Majority rule now performs a function: it indicates (with some chance of error) whether the initial string had more ones than zeros.

8.4 The Edge of Chaos

The idea of "the edge of chaos" originated with a few simple computational experiments by Packard (1988b) and Langton (1990) and quickly became part of the lexicon of complex systems. These early experiments suggested that systems poised at the edge of chaos had the capacity for emergent computation. The intuition behind this claim has tremendous appeal: systems that are too simple are static and those that are too active are chaotic, and thus it is only on the edge between these two behaviors where a system can undertake productive activity. In its most grand incarnation, the edge of chaos captures the essence of all interesting adaptive systems as they evolve to this boundary between stable order and unstable chaos.

The notion of the edge of chaos is both intuitively appealing and metaphorically rich. While the examples of Packard and Langton are intriguing—and scientific progress often proceeds from examples, to ideas, to understanding—they may also be misleading, lacking adequate logical underpinnings. That said, there remains a grain of truth in the basic intuition that, loosely speaking, complexity lies somewhere between order and chaos.

Here we explore the edge of chaos in a simplified framework. Our goal is to begin to solidify the boundaries of our understanding of this concept through a more formal analysis. We investigate two questions: is there an edge and, if so, is it important? If the idea does have merit, then its impact on our understanding of complex adaptive social systems could be substantial.

8.4.1 Is There an Edge?

A key first step in understanding whether there is an edge of chaos is being careful about defining the space that we are exploring. In models of the edge of chaos, there is an attempt to detect whether *a given rule* will tend to imply a system that is either chaotic, stable, or poised somewhere in between. A casual reading of this statement (without the added emphasis) often causes the misleading perception that the focus here is on the implied phase space of the rule rather than the rule itself.

The "edge" in the edge of chaos is not in phase space but in the space of rules. The idea is that if we slightly perturb a rule that generates complexity we will get a rule that either generates chaos or stasis. Therefore, the search for the edge of chaos focuses on how small changes in a rule impact its behavior.

We discuss the edge of chaos in terms of a one-dimensional, two-state, nearest-neighbor cellular automata. In such a system, a rule lies at the edge of chaos if small changes in its rule table move it back and forth between chaotic and nonchaotic behavior.

Our investigation requires us to classify each rule according to the behavior it generates. As previously discussed, Wolfram (2002) classifies automata as either being fixed (Class 1), periodic (Class 2), chaotic (Class 3), or complex (Class 4). Li and Packard (1990) use a slightly different and finer classification. For our purposes, the important difference between these two classifications is that Li and Packard pay special attention to rules that generate two-period cycles, what we will call *blinkers*. Li and Packard's definition of complex is also worth noting. Like Wolfram, they define a rule as complex if the time it takes to get to the limiting distribution is long, but in addition they require that this time increases linearly with the number of cells.

TABLE 8.11
Rule 110

Situation	Left Neighbor	Self	Right Neighbor	Rule 110
0	0	0	0	0
1	0	0	1	1
2	0	1	0	1
3	0	1	1	1
4	1	0	0	0
5	1	0	1	1
6	1	1	0	1
7	1	1	1	0

TABLE 8.12
Neighbors of Rule 110 and Their Respective Wolfram Class and Li-Packard Classifications

Situation	Rule 111	Rule 108	Rule 106	Rule 102	Rule 126	Rule 78	Rule 46	Rule 228
0 (000)	1	0	0	0	0	0	0	0
1 (001)	1	0	1	1	1	1	1	1
2 (010)	1	1	0	1	1	1	1	1
3 (011)	1	1	1	0	1	1	1	1
4 (100)	0	0	0	0	1	0	0	0
5 (101)	1	1	1	1	1	0	1	1
6 (110)	1	1	1	1	1	1	0	1
7 (111)	0	0	0	0	0	0	0	1
λ	3/4	1/2	1/2	1/2	3/4	1/2	1/2	3/4
Wolfram	2	2	3	3	3	1	2	1
Li-Packard[a]	2-C	2-C	Ch	Ch	Ch	F	F	F

Note: [a] 2-C is 2-cycle, Ch is chaotic, and F is fixed.

Consider Rule 110 shown in table 8.11. This rule is classified as complex by both Wolfram and Li and Packard. To see whether it is poised at the edge of chaos, we need to define the set of neighboring rules. The most obvious notion of neighborhood here is to consider all the rules that have a rule table that differs by only one situation. Using this convention, Rule 110 has the eight neighboring rules given in table 8.12.

Is Rule 110 at the edge of chaos? The last two rows of table 8.12 reveal that three of its eight neighbors are chaotic[11] and three lead to

[11] The usual caveats of rule classification apply; for example, all three chaotic neighbors have fixed points at all zeros.

fixed points. Thus, this rule appears to be poised between chaos and stasis. Superficially, at least, this seems to indicate that Rule 110 is on the edge of chaos.

Langton's (1990) model of the edge of chaos classifies one-dimensional cellular automata according to a single parameter, λ. In his model each site has s possible states and is connected to k neighbors in both directions. The λ value for a given rule equals the percentage of all rule table entries that map into some predefined *quiescent* state. Thus, if all rule table entries map into the quiescent state, then λ will equal one. In this case, the system will immediately freeze in the quiescent state. In his experiments, Langton explored randomly generated rule tables and tried to connect the λ value for a given rule table to a measure of subsequent system activity. He found that λ had some explanatory value—as it decreased from one, the average behavior of the implied systems went from rapidly freezing, to long transients, to chaos. For Langton, the edge of chaos was the value of λ at which the average behavior first showed evidence of chaos.

It is not too surprising that chaotic system behavior roughly correlates with a parameter like λ. As a crude approximation, suppose that there are only two possible states and that at any time step the state of each site is randomly chosen.[12] With random sites, the λ parameter gives the probability that any given site will be mapped to the quiescent state. Thus, when λ approaches either of its extreme values (zero or one), the sites are likely to lock into the quiescent state. As λ gets closer to $1/2$, chaos will reign as each site will have an equal chance of being either value. Thus, an "edge" associated with monotonic changes in λ will appear between these two types of behaviors.

This crude approximation of what occurs as λ varies can be refined by looking at particular rules in more detail. Our approach here is to explore the entire space of two-state, nearest-neighbor cellular automata rules. Of the 256 rules of this type, 32 are classified as chaotic. Since there are eight rule table entries for this type of cellular automaton, λ values are of the form $j/8$, where j belongs to $\{0, 1, \ldots, 8\}$. The number of rules associated with each λ value varies widely, from 1 rule for λ equal to 0 or 1, to 70 rules when λ equals $1/2$. In table 8.13 we show the number of possible rules for each λ and the number of such rules that are classified as chaotic and complex.[13] The table shows that the chaotic rules are strongly biased toward the middle of the distribution of λ (with

---

[12] We know, of course, that as time passes there is often much more structure to the sites (for example, triangles and such), but for the moment we ignore this issue.

[13] The chaotic rules are: 18, 22, 30, 45, 60, 73, 75, 86, 89, 90, 101, 102, 105, 106, 109, 120, 122, 126, 129, 135, 146, 149, 150, 151, 153, 161, 165, 169, 182, 183, 195, and 225. The complex rules are: 54, 110, 124, 137, 147, and 193.

TABLE 8.13
λ-Distribution over Chaotic and Complex Rules in the Space of Two-State, Nearest-Neighbor, One-Dimensional Cellular Automata

λ	All Rules	Chaotic Rules	Complex Rules
0	1	0	0
1/8	8	0	0
1/4	28	2	0
3/8	56	4	1
1/2	70	20	4
5/8	56	4	1
3/4	28	2	0
7/8	8	0	0
1	1	0	0

TABLE 8.14
$\hat{\lambda}$-Distribution over Chaotic and Complex Rules

$\hat{\lambda}$	All Rules	Chaotic Rules	Complex Rules
0	2	0	0
1/8	16	0	0
1/4	56	4	0
3/8	112	8	2
1/2	80	20	4

most having a λ value of 1/2). The table also shows the distribution of the six complex rules in this space. As expected, these rules too are biased toward the middle of the λ distribution.

Given symmetry, both zero and one can be thought of as quiescent states. Therefore, a λ of 5/8 and one of 3/8 are equivalent, and so on. If we define $\hat{\lambda}$ to be the minimum of λ and $(1 - \lambda)$, we can condense table 8.13 into table 8.14. Using these data we can compute the average $\hat{\lambda}$ for the different types of rules. Chaotic rules have an average $\hat{\lambda}$ of 0.44 and complex rules an average of 0.46. Thus, Langton's inference, namely that complexity occurs at the edge of chaos, seems to hold in this example.

However, the edge that appears in the aggregate data is not apparent in the individual cases. That is, there are multiple edges, not just a single one. For example, consider the neighbors of Rule 110 (see table 8.12). If the picture of an easily tuned world existing between complexity on the one hand and chaos or stasis on the other was accurate, then we would expect that the neighbors with λ equal to 1/2 would generate chaos and those with λ equal to 3/4 would be cyclic or stable. Yet, we see from

TABLE 8.15
Relevant Rule Table for Rule 46 after Two Iterations

Situation	Rule 46
0 (000)	0
1 (001)	1
2 (010)	–
3 (011)	1
4 (100)	0
5 (101)	1
6 (110)	0
7 (111)	–

table 8.12 that this is not the case. When λ equals 1/2, two rules are chaotic, two are fixed, and one is cyclic. When λ equals 3/4, one rule with each type of behavior exists.

To investigate this issue a bit more, consider Rule 126 shown in table 8.12. This rule has a λ value equal to 3/4 yet it is chaotic. A look at its rule table indicates that it produces a one unless the site and its neighbors all agree, in which case it goes to zero. While this rule could potentially lock into all zeros, its general tendency is to propagate ones in the system while eating away at the edges of any long strings of zeros. Moreover, as soon as three consecutive ones appear, they are destroyed by having a zero inserted at their center. As long as the system begins at any configuration other than all ones or all zeros, Rule 126 tends to create ones and then destroys them, resulting in long, convoluted cycles of activity.

Now consider Rule 46, which has periodic behavior with a λ of 1/2. Some initial intuition can be gained by considering the behavior of the rule after one time step. To do so, we can look at a neighborhood of size 5 (which has thirty-two possible configurations) and track what happens to the middle three sites. We find that the sequence 010 is not a possible outcome—that is, after one iteration of the rule we can never have a single one bordered by zeros. This means that any future configurations that require isolated ones will also be ruled out after the second iteration, and in the case of Rule 46 this rules out the sequence 111. Therefore, after two periods, we will never see the sequences 010 or 111, and we can reduce the relevant parts of the rule table to that given in table 8.15. In this restricted domain, Rule 46 is much simpler: copy the site to the right. Such a rule is Class 2 according to Wolfram and fixed based on Li and Packard—in either case, it is not chaotic.

The failure of a nicely behaved λ-edge in the neighborhood of Rule 110 is not particular to that rule. Every one of the rules classified as complex in this space has at least one chaotic neighbor with a lower λ value and one with a higher value. Therefore, although it is true that complex rules have chaotic edges, they do not lie poised at the edge of chaos in the traditional sense implied by λ. That being said, the edge of chaos idea is not wrong per se, as complex rules do appear to lie next to chaotic rules. However, collapsing a multidimensional phenomenon onto one dimension obscures the details. In the case of these simple automata, what drives complexity, chaos, and order is the microstructure within a rule such as we saw in Rule 46.

To show how microstructure undermines attempts to create a "complexity dial," we can partition the set of rules into four equal-sized groups (with sixty-four rules in each) based on how the rule maps the sequences 000 and 111. We will name each sequence based on the implied behavior when we have a string of all zeros or all ones. Let the *identity rules* be the ones that map 000 to 0 and 111 to 1. We call rules that map 000 to 1 and 111 to 0 *blinker rules*, as a string of all ones goes to all zeros, then back to all ones, and so on. The third and fourth sets are those rules for which 000 and 111 both get mapped to either 0 (*0-attractor rules*) or 1 (*1-attractor rules*). In these latter rules, a string of identical elements falls into, and remains in, the attractor state after one time period. Both identity and blinker rules have an identical distribution of the number of ones in the rule tables (with one rule each having one or seven ones, six having two or six ones, fifteen having three or five ones, and twenty having four ones). The two attractor rules have the identical distribution except for the number of ones is offset either down by one (0-attractor rules) or up by one (1-attractor rules). The λ distribution for each class is directly tied to the distribution of ones.

Table 8.16 classifies the rules within each of the four sets using Wolfram's classification. Note that the blinker rules are far more likely than the other types of rules to be periodic. This is not surprising, as once long sequences of zeros or ones emerge, these rules embody two-period cycles. Indeed, of the fifty blinker rules that lead to periodic behavior, forty-six of them frequently generate cycles of period 2.[14] Also note that the identity rules and the attractor rules tend to be far more stable then the blinker rules. Again, given the tendency of these rules to lock in long sequences of zeros or ones, this is not surprising.

The one potentially puzzling result revealed in table 8.16 is that all of the complex rules belong to the set of attractor rules. Recall that attractor rules map 000 and 111 into the identical value, and thus it would seem that these rules are the most likely candidates for boring behavior among

---

[14] In contrast, of the thirteen identity rules that are periodic, only nine generate period 2 cycles.

TABLE 8.16
Classification of Our Rule Partition into Behaviors

Rule Set	Class 1: Fixed	Class 2: Periodic	Class 3: Chaotic	Class 4: Complex
Identity	48	13	3	0
Blinker	7	50	7	0
0-Attractor	29	17	11	3
1-Attractor	29	17	11	3

the four sets. When we look at the complex rules within these sets, we find a common characteristic— namely, that their rule tables are heavily biased toward the opposite value that is being mapped to under 000 and 111. This leads to an inherent amount of, in Schumpeter's words, creative destruction. Long sequences of identical values that are induced and stabilized by either the 000 or 111 mappings are destroyed at the edges by the other elements of the rule table. During this destruction, long sequences of the opposite value begin to accumulate, and these form the basis for the creation of the original values. This process of creative destruction results in long transients, where one value is churned into another. This also explains why the identity rules do not generate complex behavior, as long sequences are stabilized rather than destroyed under such rules.

Our decomposition into the four sets also proves useful when thinking about neighbors and the edge of chaos. Six of the eight neighbors of any rule belong to the same set within the decomposition. Thus, six of the eight neighbors of an identity rule will themselves be identity rules and thereby are likely to generate fixed points. None of an identity rule's immediate neighbors can be blinker rules (as this requires flips in the rule table for both 000 and 111), and therefore stability does not appear likely to lie at the edge of periodicity. Attractor rules, which have exactly one identity-rule and one blinker-rule neighbor, tend to border attractor rules. More than two-thirds of the chaotic rules are attractor rules, and all of the complex rules fall into this set. This partly explains why complex rules have chaotic edges, as the complex and chaotic rules belong to the same set. Therefore, our decomposition (by construction) lends support to the notion of an *edge of chaos*, but not a single edge, as usually supposed, but rather a multitude of edges contained within the set of attractor rules.

A crude measure like λ has its uses and limitations. As the analysis indicates, knowing the λ value of a particular rule does not necessarily tell us much about that rule's behavior. The λ value is, however, probably a good way to identify broad areas of the rule space that might harbor the potential for interesting behavior. In this view, λ is a necessary but not sufficient condition for interesting behavior. Note that the analysis was confined to very simple automata (one-dimensional, two-state, with one-nearest-neighbor), and we know that such systems may have limitations (though sometimes even advantages) over more complicated systems. That being said, we suspect the tenor of the insights obtained here are relatively general.

8.4.2 Computation at the Edge of Chaos

For most physical, biological, and social processes, an important aspect of rule behavior is whether the system can generate productive behavior. The productive behavior we are concerned with here is the ability of a system to solve a computational problem. The measures of system behavior we used were ways to capture indirectly an automaton's ability to compute answers by tracking its ability to transmit and process information. In this section we look at the process of computation more directly.

Cellular automata become computational systems when they produce "answers" to "questions." In these systems the questions are posed by setting up initial conditions and then activating the automata. Answers come from some interpretation of the system after it has undergone sufficient iterations. For example, if the system must determine a binary answer to some question, we might allow the automaton to run and then take the state of a randomly chosen site as the answer. A well-known example of such a computation (Packard, 1988b; Mitchell, Crutchfield, and Hraber, 1994) is having the automaton determine whether the initial condition has more ones than zeros.

Because computations must give useful answers, intuitively we would suspect that the rules that can undertake computation must lie in a regime that is neither too chaotic (since no consistent answers will be forthcoming) nor too ordered (since insufficient computation can take place). If we rely on a measure like λ, we may not have strong support for this proposition, since experiments like Mitchell, Crutchfield, and Hraber (1994) show that useful computation takes place where λ is much closer to 1/2 than one would predict.

To investigate these issues, consider automata that must solve binary classification problems. These automata take some initial condition and classify it into one of two possible answers (say, yes or no). We will allow an automaton to take any given question (initial condition) and compute until it enters its limit cycle, at which point we will query a random site at a random iteration to get the answer. A computation will yield the wrong answer if the queried site is in the wrong state. For an automaton to yield a *perfect computation*, it must enter a one-period limit cycle with every site having the identical, correct answer.

An initial step in analyzing such a system is considering only those automata that yield perfect computations. The requirement of perfect computation is quite severe as there are many situations in which imperfect computations, say, those that give the correct answer 85 percent of the time, could be of great value. Nonetheless, starting at perfection allows us to narrow the analysis sufficiently and develop some useful benchmarks.

A necessary condition for perfect computation is that the automaton enters a one-period limit cycle in which all sites are in the same state. Thus, for a perfect computation the rule table must always map 000 into a 0 and 111 into a 1. This constraint is the defining property of the *identity* rules discussed previously.

At the outset, we note that these kinds of systems are not going to be able to solve that many problems. Since each automaton is deterministic, any given question (initial condition) will always be associated with the same answer (either a one or zero). Thus, each automaton can only implement a single classification scheme. Recall that there are sixty-four possible identity rules, and therefore at most we will be able to do sixty-four unique computations.[15] More likely than not, some automata will overlap and solve the identical problem. Thus, even under the best of circumstances, there can be at most sixty-four unique computations (binary classification schemes) possible. That is, these kinds of decentralized computational systems, with one-dimensional, two-state, one-nearest-neighbor automata, can solve perfectly no more than sixty-four different kinds of classification problems.

The mapping restriction that we placed is a necessary, but not a sufficient, condition for perfect computation. All it does is ensure that if the automaton can configure itself such that all sites have the right answer, then it can maintain that answer. The restriction does not eliminate other limit cycles that would result in imperfect answers.

What about the other types of rule classes? Blinker rules, given our requirement of stabilizing on a single answer, will not work as they alternate between ones and zeros. Attractor rules might be able to classify in one direction. That is, they can produce all zeros (or all ones) if a criterion is met and not converge otherwise.

---

[15] If a given rule is capable of universal computation and if we allow customized input strings, then we could perform a vast array of computations. Of course, this requires "intelligence," also known as computation, being embodied in the initial input string. As can be seen from table 8.16, there are no identity rules that are complex. This is an artifact of the size of the neighborhood, as rules with more neighbors can belong to the class of identity rules and be complex, as Mitchell, Crutchfield, and Hraber (1994) show.

These simple distinctions between the various classes of rules allow us to speculate on how systems could evolve the capacity to compute.

First, note that decentralized computation becomes much more difficult if the initial conditions leading to different answers are "close" to one another. Therefore, successful early computation may require the appropriate environment—one in which the "answers" are "obvious." Once the system achieves this milestone, it can start to fine-tune itself to handle more difficult cases. This secondary tuning of the algorithm probably does not respect crude boundaries based on statistical generalizations like the λ parameter, any more than, say, height is an indicator of basketball skill. Thus, in systems such as the one studied by Mitchell, Crutchfield, and Hraber (1994), we are likely to find that gross measures of system behavior, like λ, are insightful as the initial evolution embraces the easy cases. As the system further evolves to handle the more difficult cases, however, measures like λ become less relevant.

Second, given our rule classes we might expect that remedial computations first emerge via identity-type rules. Once these computations become available, the other classes of rules may come into play. To put this more concretely, early on blinkers and attractors may get selected against in favor of identity rules. Once identity rules are established, we might expect that attractor rules will arise. Attractor rules are capable of more interesting computation than either blinker or identity rules and are more likely to emerge from identity rules as they require only a single mutation in an identity rule's table (versus two for a blinker rule).

The fact that we find little evidence that a one-dimensional characterization of the edge of chaos suffices as an accurate indicator of computational ability could well be an artifact of our smaller-sized automata, but a more likely hypothesis is that the metric needs to be better situated within the space of automata rules. At the simplest level, rule structure is a function of both the number of zeros in the rule table and the location of these zeros. There is a delicate dance between these two features when we are away from the extremes. If we want to understand these types of systems, we must be willing to disentangle their movements.

8.4.3 The Edge of Robustness

We end this exploration into the edge of chaos on a more speculative note. Fine-tuning these systems creates a tension. As we attempt to incorporate more delicate behavior by adding more structure to a rule, we are likely to make the underlying system less robust. This is because the structures necessary for delicate behavior require an underlying system that is rich in possibilities. In essence, we need a quivering system that will fall into the right state with only a gentle tap. In such a system, an improper tap can lead to very unpredictable results.

Adaptive systems have to deal with the tension between the benefits of achieving precise behavior and the cost of increased system fragility. One hypothesis is that adaptive systems will have a bias toward emphasizing simple structures that resist chaos over more complicated ones that handle difficult situations. There are two reasons for this hypothesis. The first is that simple structures are likely to be easier to find and maintain. Of course, that does not eliminate the possibility that adaptive systems first lock on to these simple structures and then move on to more complicated ones. This is a plausible supposition, if we assume that there is an adaptive path from one to the other.

The second justification for the hypothesis is that systems that are fragile are very risky in terms of rewards, and adaptive systems tend to be risk averse. While being able to handle delicate situations appropriately on occasion might result in large rewards, there is also a chance that it will lead to large losses. Adaptive systems tend to be inherently risk averse because, notwithstanding the potential gains to be made by taking even a favorable risk, it takes only a single loss to kill off an agent and eliminate it from the system forever.

It is hard to disagree with the notion that adaptive systems will tend to evolve agents with behavior that is between less-productive extremes. Such a weak-form edge-of-chaos hypothesis seems both sound and a useful starting point from which to launch further investigations. The strong-form hypothesis—namely, that adaptive systems congregate at a narrow edge where slight changes in their behavior lead to chaos or frigidity—is harder to justify. At least in the case of our simple automata, rules that compute lie in a region of rule space that consists largely of other rules that compute.

The full implication of these edge-of-chaos hypotheses for the social sciences is still an open question. Clearly there are systems, for example, stock markets, in which agents actively adapt and alter the fundamental behavior of the system and, in so doing, force it into new realms of activity. Thus, if stock markets are too predictable, then we would expect adaptation to create agents that can exploit this feature. The emergence of such agents should wipe out the predictability and push the system toward a more chaotic regime (which is essentially the argument driving the efficient-market hypothesis). However, once the market is completely chaotic, the selective pressures on agent behavior become quite neutral, and predictability might again slip back into the system.

CHAPTER 9

Social Dynamics

> Far better an approximate answer to the right question,
> which is often vague, than the exact answer to the wrong
> question, which can always be made precise.
> —*John Tukey, Annals of Mathematical Statistics*

> Few things are harder to put up with than the annoyance of a
> good example
> —*Mark Twain, The Tragedy of Pudd'nhead Wilson*

To FURTHER OUR INVESTIGATION of complex adaptive social systems,
here we create some models with more elaborate agent dynamics. These
dynamics allow us to investigate new realms of social behavior, and the
resulting models can be used to explore topics such as racial segregation,
the role of expectations on behavior, and city formation. Moreover, we
also consider some new concepts surrounding equilibrium analysis and
self-organization in social systems.

9.1 A ROVING AGENT

Our first model considers a system composed of a single agent who
maneuvers in physical space. A single agent does not a society make,
so to transform this into a model of a social process we would need to
tell some story about how influences from other agents in the world are
embodied in our single agent's movement rule. Creating a coherent and
socially meaningful story along these lines is not difficult, but we will
forgo such opportunities as our focus here is to illustrate some higher-
level issues surrounding equilibria.

Our single agent has an odd penchant for roving across a line with the
locations numbered sequentially from 0 to N. In each period the agent,
using a fixed-movement rule, decides on a new location based only upon
its current location. The movement rule depends on whether the agent
finds itself in either the lower- or upper-half of the line. If it is in the lower
half, that is, at site j where $j \leq N/2$, it moves to site $2j$; if, instead, it is
currently located in the upper half ($j > N/2$), it moves to site $2(N - j)$.
Given this rule, an agent in the lower half always increases its location

TABLE 9.1
Rover Dynamics for $N = 10$

Location	Next Location
0	0
1	2
2	4
3	6
4	8
5	10
6	8
7	6
8	4
9	2
10	0

and eventually enters the upper half of the line, unless it is at location 0, in which case it stays there forever. An agent in the upper half will go to the lower half when $j \geq (3/4)N$ and stay in the upper half when $j < (3/4)N$. When $j = (2/3)N$, the agent will stay exactly where it is. Thus, the system has two fixed points, one at 0 and one at $(2/3)N$. Table 9.1 summarizes these dynamics for the case where $N = 10$.

We can use some simple computational experiments to understand better the potential dynamics of the roving model. Recall that earlier we showed how finite systems with deterministic dynamics (like this one) are guaranteed to cycle, so our analysis here focuses on these cycles. Consider systems where the agent always begins at site 1. If the line has ten sites, then from table 9.1 we see that the agent will find itself in the following locations: 1, 2, 4, 8, 4, ... Note that once it returns to a site previously visited (site 4 in this case), it begins to cycle. With one hundred sites, the agent will consecutively locate at sites 1, 2, 4, 8, 16, 32, 64, 72, 56, 88, 24, 48, 96, 8, ..., and thus be ultimately trapped in a cycle of length ten that perpetually traverses the values between 8 and 96 in the above series. In table 9.2 we show the cycle lengths for other values of N. As is apparent from the table, as N increases so does the cycle length.

The function that underlies the behavior of our agent is a discrete analog to the "tent map" made famous in chaos theory (see, for example, Ott, 1993). In the limit, the cycles become infinitely long, implying that our agent will forever roam the lattice.

In the continuous case defined over the open interval, the tent map has a unique, fixed-point equilibrium at $x = 2/3$. In the discrete examples that we just considered, N was not divisible by 3, so this equilibrium did not exist. We can fix this problem by letting $N = 102$, in which case a

TABLE 9.2
Equilibrium Cycle Length for a Single, Roving Agent

Number of Sites	Cycle Length
10	2
100	10
1,000	50
10,000	250
100,000	1,250

fixed point arises at location 68 (since $68 = 2(102 - 68)$). Suppose that an agent at the fixed point makes a slight error and lands on location 67. From location 67, the agent would next travel to locations 70, 64, 76, 52, 100, 4, 8, 16, 32, 64, ..., ending up in a cycle of length eight. Thus, this equilibrium point is unstable. This instability is tied to the steep slope of the relocation function. Namely, small mistakes lead to larger corrections—after moving one location too low, the agent moves two locations too high, then four locations too low, and so on.

While the lack of stability of the fixed-point equilibrium is problematic, of even more concern is its relatively small basin of attraction. If we assume no mistakes, an agent will end up at location 68 only if it starts at locations 17, 34, 85, or 68. Similarly, for the fixed point located at 0, only locations 51, 102, and 0 will get you to that equilibrium. Thus, only a small fraction of the total locations will get you to the fixed point, implying that such equilibria may be hard to acquire (and, with noise, easy to lose).

In this simple model the traditional notions of fixed-point equilibrium analysis may not be that useful. While the model is guaranteed to fall into an equilibrium, the cyclic equilibria have much larger basins of attraction than the fixed-points ones. The combination of multiple equilibria, small basins of attraction for the fixed points, and instability suggests that traditional equilibrium analysis may have little predictive value. Complex systems models may enable us to identify which equilibria are most likely and help to reveal the links between behavioral assumptions and equilibrium selection.

9.2 SEGREGATION

A classic "computational" social model is Schelling's (1978) work on racial segregation. In this model, the world consists of agents living on the squares of a checkerboard. Initially, agents are randomly sprinkled across the board (with one agent per square and some squares left unoccupied). There are two types of agents, and each agent has a minimum required threshold for living with those of the same type. When the number of same-type neighbors falls short of an agent's threshold, it randomly relocates to a new spot on the board. One of this model's most striking results is that even when agents are tolerant of the opposite type, segregation is still likely to emerge. Segregation arises due to the phenomenon of tipping, whereby the early movements of even a few agents can create the incentive for other agents to move. This cascade of movement dies out only when the system becomes highly segregated.

We now create a one-dimensional version of Schelling's Segregation model. Assume that two types of agents live on a circular lattice and that each agent's neighborhood consists of all locations k steps away. Each site on the lattice is initialized with either an agent of one of the two types or it is left empty (with equal probability among the three possibilities). Following Schelling, each type of agent has the identical threshold value, but these values may differ between the two types. When an agent is given a chance to act, it first calculates the percentage of its neighbors that are of the same type. If this value is less than its threshold, the agent moves to a random, vacant location; otherwise, it remains in place. We assume a location-based, asynchronous updating mechanism—in each period we loop through the sites in spatial order and allow the occupant of the site, if any, to act.

To analyze our model, we distinguish between tipping and segregation. We define tipping as the process by which the movement of agents causes cascades of further movement. We operationalize tipping by comparing the number of agents that move before the model reaches an equilibrium to a measure of the potential for agent movement (either the number of agents that want to move initially or the number that actually move during the first iteration[1]). Increased tipping is associated with higher values of the ratio of the number that actually move to the potential number of movers. Segregation occurs when the two types of agents fail to associate with one another. We measure segregation by counting the number of times two neighbors (ignoring any intervening spaces) are of different types. Lower values of this measure are associated with greater segregation in the system.

We run the model on a lattice of size 100 with a neighborhood size of four. The data is averaged over 500 separate trials.

In the initial model we give each type of agent a threshold of 40 percent (the 40/40 column in table 9.3), whereby the agent will move if fewer than 40 percent of its neighbors are the same type. Under this condition

---

[1]Because agents are acting asynchronously, the number that actually moves during the first iteration may differ from the number that initially wanted to move.

TABLE 9.3
Results from a One-Dimensional Tipping Model with $N = 100$, $k = 4$, and 500 Trials

	40/40	30/50	31/49
Agents initially wanting to move	19.2	26.0	19.8
Agents who move during first iteration	21.9	30.1	23.2
Agents who move during first 100 iterations	42.5	110.8	58.8
Final blocks	5.8	5.0	5.8

there is not much tipping—the number of agents that eventually move is only about twice that of those who wanted to move initially. Despite the low amount of tipping, segregation increases substantially, with the number of opposite-type adjoining neighbors decreasing from around 30 to 5.8.

These results are a bit puzzling as the amount of tipping is well below that observed by Schelling. A number of differences in our model might account for this discrepancy: we use only one versus two dimensions, our updating rule is slightly different, and threshold choices may be important. Additional experiments suggest that while the dimensionality and updating can make a difference, the more interesting area of investigation is in the choice of thresholds.

One of the major differences between the parameters of this model and Schelling's is that we used symmetric thresholds. Schelling relied on asymmetric thresholds for the two types of agents. The 30/50 column in table 9.3 gives the result when the two types of agents have 30 percent and 50 percent thresholds, respectively. Under these thresholds, tipping becomes much more pronounced, with around four times the number of agents that initially wished to move eventually moving. Although the amount of tipping is much higher, the final segregation level is roughly equal to that seen under the symmetric threshold.

Given the discrete nature of the neighborhood, there is the potential for discontinuities to arise in the the model. These discontinuities can become particularly important under asymmetric thresholds. Because agents base their actions on the *percentage* of neighbors, the critical thresholds are tied to the neighborhood size. With only two neighbors, you can have only 0 percent, 50 percent, or 100 percent of the same neighbor type; with three neighbors, you can have 0 percent, 33 percent, 67 percent, or 100 percent; with four, you can have 0 percent, 25 percent, 50 percent, 75 percent, and 100 percent; and so on. These discontinuities imply that the behavioral differences induced by various thresholds may be slight. For example, with anywhere between two and eight neighbors, 40 percent versus 50 percent thresholds only imply different behavior in two out of the forty-two possible configurations of the world. To demonstrate such effects, we ran the model with asymmetric thresholds of 31 and 49 percent. As shown in table 9.3 these slight changes in threshold values cause the system to resemble the symmetric 40 percent thresholds case rather than the parametrically much closer asymmetric case.

Although the data indicate that the final segregation level is similar across the parameters, the dynamics are very different. With symmetric thresholds, the system rapidly converges to a segregated outcome as agents of both types find acceptable neighborhoods. With asymmetric thresholds, segregation takes much longer. In these systems, almost all of the movement is by the agents with the less tolerant threshold. On average, twenty-three of the twenty-nine agents that moved in the first period had the less-tolerant threshold. This behavioral discrepancy continues on into later periods as well. This differential movement explains the puzzle of why there is so much more movement, yet similar segregation, in the asymmetric system: with only one type of agent moving, it takes longer for the system to segregate.

Finally, we can further refine our analysis of movement by tracing the impact of a given move on subsequent moves. One possibility here is that when an agent moves, it is still unhappy with its new location and moves again. This type of movement is a common occurrence with a 50 percent threshold. Another possibility is that the agent's move impacts either its old neighborhood (where neighbors of the same type now want to move) or its new neighborhood (where agents of the opposite type now want to move). We find that in the asymmetric case, there tends to be far more of the former type of tips—that is, agents in the old neighborhood decide to move when like-type neighbors leave. In sum, depending on our assumptions we can alter the amount of tipping. We find that high levels of tipping are associated with the 50 percent rule. Relocations in a Segregation model are a form of positive feedback, and when conditions are right such feedback can cause major cascades to wash across the system.

9.3 The Beach Problem

Though less well known than his Segregation model, Schelling's (1978) Beach model has become a core model of complex adaptive social systems, albeit under different names.[2] In Schelling's version, every

---

[2]Arthur's (1994) variant is known as the El Farol problem. In physics there is a version called the Minority game, in which an odd number of players chooses either option A or B, and those who choose the minority option split a prize (Challet and Zhang, 1997).

Saturday people living in the city must decide whether to go to the beach. While everyone enjoys getting away to an uncrowded beach, their enjoyment is greatly curtailed if the beach is crowded.

A few obvious social mechanisms could solve such a problem. One method would be to impose some kind of institutional structure from the top down that restricts access to the beach. This could be in the form of a market (you need to buy a ticket to get on the beach) or some other allocative arrangement like, say, allowing admission based on some immutable characteristic like the parity of the beach goer's license plate.

Alternatively, we could consider decentralized solutions that emerge from the bottom up, that is, we can hope that purposive individual behavior aggregates into a sensible solution. For example, game theory provides a simple bottom-up solution via mixed strategies. If the city contains a million people but only one hundred thousand can comfortably coexist on the beach, then each person can write the word "beach" on one side of a fair, ten-sided die, and only go to the beach if a roll of the die yields "beach." In this way, on average, the optimal number of people will be at the beach each Saturday.

This simple mixed-strategy solution has two potential flaws. The first is relatively minor, namely, that while on average one hundred thousand people will go to the beach, there is some variance across all of the dice throws, and the number going to the beach will rarely be exactly one hundred thousand. It turns out that the standard deviation of this random process is only about three hundred, implying that more than 99 percent of the time we would not expect the actual number on the beach to deviate by more than one thousand people. Thus, sometimes the beach is a little crowded and sometimes there is a bit of extra space to stretch out, but overall randomness does not seem to be much of an issue here.

The second, and more problematic, issue is that we rarely see individuals employing dice in such decisions. More typically, people base their actions on various predictive models of the world, and then the aggregation of these predictions drives the system. These predictive models require individuals to make decisions based on recognizable patterns that form in the world. Presumably, individuals will tend to latch on to those models that preform better, that is, that result in behavior that leads to superior outcomes. Of course, there is an infinity of patterns one could look for in the world, and each of these can be associated with various decisions. For example, if the beach was crowded last Saturday (or, say, three out of the last five Saturdays), then one person might reason that it will be empty next week, while another might believe that it will remain crowded. Whatever the rules, we are likely to see heterogeneity among the population. Moreover, these various predictive models will interact with, and adapt to, one another much like organisms in a biological ecosystem. The models will compete with one another, alter, form new niches, and so on in a very dynamic process.

Unpacking the implications of such a formulation was the focus of Brian Arthur's (1994) reformulation of the Beach problem as the El Farol problem. In this version, people want to go to a local Santa Fe bar. Like Schelling's Beach model, they would rather stay at home if the bar is crowded. Arthur constructed a model in which each agent had access to several decision rules that competed with one another for the agent's attention (based on the rule's past predictive value). Arthur found that the outcome of this adaptive system was quite efficient, in the sense that the number of people who showed up in the bar each weekend was, on average, optimal.

In subsequent work (mostly by physicists), many variations have been explored. It has been found, for example, that the model can produce less variance than would exist if each person had rolled a die and played the mixed strategy. This reduction in variance occurs even though different people go to the bar each week. Thus, decentralized, diverse agents can self-organize so as to create reasonably stable outcomes (Zambrano, 2004).

We can illustrate some of these ideas using a simple model. Suppose we have four people in the world, A, B, C, and D, who want to get together for some activity once a week. The activity is such that it is great fun if only two show up, but if any other number arrives it would have been better to have just stayed at home. We further assume that everyone knows how many people showed up each week. To seed the system, let all four people show up to the first meeting and only A and B to the second.

We begin our analysis by assigning each person a simple decision rule. (These rules were derived somewhat arbitrarily using some commonsense prototypes.) Person A goes if the *average* attendance over the past two weeks is less than two. Person B goes unless exactly two people went last week. Person C goes if more people went last week than the week before. Finally, person D goes if exactly two people went last week.

By applying these rules we can determine the weekly attendance patterns. Thus, in week 3, person A stays home since the average attendance has been three, person B stays home since exactly two went last week, person C stays home since the recent attendance has fallen, and person D goes because exactly two went last week. Table 9.4 shows the attendance patterns for the first ten weeks. Notice that we have the identical attendance patterns in weeks 8 and 9 that we had in 4 and 5, and

TABLE 9.4
Attendance Patterns in a Simple Beach Problem

Week	Attendance
1	ABCD
2	AB
3	D
4	AB
5	ACD
6	BC
7	D
8	AB
9	ACD
10	BC

therefore (since any given decision rule relies on at most the previous two weeks) the system will cycle *AB*, *ACD*, *BC*, and *D* forever. On average, this cycle has exactly two people showing up each week with a variance equal to one-half. If these agents used a mixed-strategy solution and each flipped a fair coin to decide if she would attend, we would achieve the same mean but the variance would increase to one. Thus, the predictive ecology approach leads to a better outcome than the mixed-strategy one in the sense that we achieve the same mean with a lower variance.

Of course, we did choose which predictive rules to admit to the ecology, so perhaps this outcome is anomalous. Alas, the rules we picked were pretty much random choices (we promise), so we did not intentionally build anything into the system. Nonetheless, there are a lot of rules we could have used, so perhaps we just got lucky.

Once we allow the possibility of agents adapting their rules we add an important twist to the dynamics. Suppose that the pattern that emerges is, say, *ABCD*, *A*, *ABCD*, *A*, and so on. Given this, we would expect at least one of the people to alter her attendance rule as the outcome is lousy for all. Presumably, this adaptation would continue until the system achieves a pattern in which each person feels that she can do no better by changing her rule.

The dynamics of such behavior can become quite complicated, and the system may or may not settle down. Consider the pattern that arose in table 9.4. Given this pattern, we might expect someone to alter her behavior. For example, suppose person *A* alters her rule so that she goes if the average attendance over the past two weeks is less than two or if exactly three people went the week before last. Table 9.5 shows the outcomes resulting from this change in person *A*'s rule. Here a new cycle emerges that takes the system through the states *AD*, *D*, *AB*, *ACD*,

TABLE 9.5
Attendance Patterns after Person *A* Changes Her Rule

Week	Attendance
9	ACD
10	BC
11	AD
12	D
13	AB
14	ACD
15	BC
16	AD
17	D
18	AB
19	ACD
20	BC

and *BC*. Again, we find that the average attendance is two, though now 60 percent of the time exactly two people show up (versus 50 percent in the previous cycle) and the new variance is even smaller than before, implying that they have found an even better global pattern.

With larger populations of agents, ecologies of predictive models become much more complicated. Nevertheless, similar ideas apply, and again we find the possibility of collections of adaptive agents creating relatively stable outcomes from the bottom up.

For example, consider the Minority game. In this game we have an odd number of players, each of whom must choose either option *A* or *B*. At each round, any player who chose the minority option shares in a prize.

In a common formulation of the game, all of the players track the minority outcome for the past *M* periods. A strategy in this game must map each of the 2^M possible histories to a choice of either option *A* or *B*. Typically these games begin with *N* players, each of whom randomly picks a strategy. Every *T* periods the worst-performing player is replaced with either a clone of the best-performing player or, with probability ϵ, a randomly selected strategy. This cloning error provides a balance between exploitation (using the best-known strategy to date) and exploration (searching for possibly better strategies).

It has been found that the ratio of the number of possible histories to the number of players is a key parameter in these models (Challet and Zhang, 1998). If $\rho = \frac{2^M}{N}$ is too small, then there is a lack of strategic diversity and efficient coordination is difficult to achieve. If ρ is large, the strategy space offers too much room within which to wander, and again efficient coordination fails. Thus, there is a critical value of

ρ that minimizes the variance of the number of agents participating in the minority. These results reveal a connection—possibly a deep one—between the depth of the cognitive models of the agents and the ability of the system to coordinate. This raises the possibility of emergent cognitive depth: adaptive agents may learn to think just deeply enough to promote coordination in the system.

9.4 City Formation

The interrelationship between economics and geography has been of interest to scholars since von Thunen's work in the early 1800s. Most recently, the impact of cities on economic growth and development has become a central topic prompted by the writings of Jacobs (1984) and others.

Most theorizing about city formation has emphasized two key aspects of the problem: space and agglomeration. A city's spatial location determines many of its resource possibilities, ranging from the price of natural resources to the ease of transportation. Furthermore, as agents agglomerate within a city, both positive and negative externalities accrue. Agglomeration effects are often subject to nonlinear feedbacks and chance—small events can have big impacts on the eventual outcome.[3] Given the emphasis of these models on spatial and nonlinear feedback mechanisms, computational tools are a natural means by which to investigate this topic.

To model the formation of cities, we construct a one-dimensional world in which each site is capable of holding an unlimited number of agents. We begin by assuming that these locations are arranged in a line. A city in such a model will be a site that is occupied by a relatively large number of agents. Agents move in response to economic and social variables associated with each site, such as wage rates, living conditions, and commodity prices. Rather than separately model each of these variables, we collapse them into two generic categories that indirectly serve as proxies for the key elements driving locational choice. The two proxies we use are the agent's home population—that is, the number of other agents located at the agent's current site—and the average distance of the agent to all of the other agents, given by $\sum_i p_i d_i / \sum_i p_i$, where p_i is the population at location i and d_i is the distance from the agent's current location to site i.

---

[3] While examples abound, the first president of the Santa Fe Institute, George Cowan, notes that Albuquerque bankers rejected a loan application from a young Bill Gates during the early years of the company that eventually became Microsoft. Gates received an alternative loan from his father, contingent on him returning to Seattle.

These two attributes, home population and average distance, capture in broad strokes the main drivers of location decisions. The home population variable allows us to manipulate how the local effects of agent agglomeration influence agent happiness. Agents may enjoy living in larger cities because of the increased city services and the employment opportunities they provide or, alternatively, find such size unpleasant due to crowding. The average distance variable accounts for the impact of more distant agglomerations on agent utility. Agglomerations of agents in other cities influence conditions like local labor markets, commodity prices, and transportation costs.

By manipulating the form of agent preferences we can create a variety of models. Here we concentrate our analysis on three simple variants. The first two variants focus on situations where agents enjoy the company of others and either want to maximize their home population or minimize their distance to others. While in both of these variants agents "like" other agents, they attempt to achieve this goal in somewhat different ways that could lead to differences in system behavior.[4] In the third variant we consider, agents attempt to *maximize* their average distance from others.[5] This last variant provides a nice contrast to the previous two.

To define the model fully, we must specify when agents get to act and how their preferences translate into actions. We will assume that agents update asynchronously based on random order. When called on to update, an agent will consider a set of potential new locations and move to the location that best meets its objectives. We can vary the amount of information an agent has about the world by limiting the potential set of locations it evaluates to, say, only neighboring sites. If desired, we could further restrict this set to be only a single, randomly chosen site rather than all of the sites in the neighborhood. Thus, our agents could range from relatively omniscient beings that look at all possible locations before deciding where to move to ones that consider only a single, random, neighboring site. A nice intermediate case between these two extremes is allowing neighborhood searches that follow a hill-climbing algorithm that iterates the search until no further improvements are possible. For example, if the neighborhood search is of size 1, an agent would initially evaluate the sites to its left and right and, if one is better than its current

---

[4] According to the 2000 census based on counties in the United States, a move to Los Angeles County, California, would accomplish the first goal and Phelps County, Missouri, would achieve the second one.

[5] If the probability of bumping into someone is proportional to an agent's distance from others, then minimizing expected contacts is equivalent to maximizing expected distance. We note that Key West, Florida, has repeatedly tried to declare itself an independent country known as the Conch Republic since 1982.

location, move to it. Once there, the agent would evaluate the next site over and move there if it improves on the site it just entered, and so on, until no further improvements are possible.

The two models where agents "like" each other have very similar behavior. In these models, the outcome that maximizes the happiness of all the agents is the one in which they agglomerate into a single city. When agents were allowed to search the line fully, this optimal outcome always emerged. The actual location of the single city depended on the initial conditions (whatever site had the largest initial population or, in the case of a tie, the one with the first immigrant catalyzed the final city). If the search was restricted to a neighborhood, then it was possible to see multiple cities forming, each separated by a distance greater than the neighborhood size. These results are not too surprising given a priori reasoning.

However, in the model where agents "dislike" other agents and want to maximize average distance, an odd phenomenon emerges. In their quest to escape one another, the agents congregate into two cities that form at the end points of the line rather than dispersing themselves throughout the world (a result that might explain the motives of people living in New York and Los Angeles). This outcome provides another nice example of macrobehavior being at odds with micromotives.

To understand this last result, consider the following. First, assume a world in which all but one agent lives at either of the two end points. Let p_0 and p_N give the populations at the two end points, and assume that the single agent resides at site i. The average distance for this agent is given by $(p_0 i + p_N(N - i))/P = ((p_0 - p_N)i + p_N N)/P$, where P is the total population. Note that the choice of i that maximizes this distance depends only on $p_0 - p_N$. If $p_0 - p_N > 0$, the agent will want to maximize i and will move to N; if $p_0 - p_N < 0$, the agent will want to minimize i and move to 0; and if $p_0 = p_N$, all locations give the agent the same distance. With the exception of $p_0 = p_N$, we will find two stable cities forming at the end points.

Next, consider a situation where the agents are spread along the line according to some distribution. The average distance of an agent located at site i is given by $(\sum_{l<i} p_l(i - l) + \sum_{r>i} p_r(r - i))/\sum_j p_j$, where p_j gives the number of agents at location j and the first two summations give the weighted distances of the agents to the left and right, respectively. If the agent moves one site to the left (that is, i decreases by 1), then the distance to those agents on the left goes down by one while the distance to those on the right goes up by one. The agent (whether it goes left or right) also gains a distance of one from whatever other agents were located at i. Thus, the agent will want to move left if $\sum_r p_r > \sum_l p_l$ (and right if $\sum_r p_r < \sum_l p_l$), that is, the agent will want to move left if there are more agents to its right than to its left. Moreover, as the agent moves and passes by other agents, the direction of movement is reinforced and the agent is driven to the end point. Based on these arguments, we will see two cities forming at either end point.[6]

We note that the results we found hold for even more elaborate topologies (Page, 1998). We can consider two-dimensional cities where distance is measured via the number of city blocks that must be traversed to get from one spot to another. Alternatively, we can array the agents around a circle rather than a line. While the proofs in these cases are a bit more elaborate, similar logic applies.

Thus, we see that a variety of micromotives, ranging from loving others to hating them, leads to the same macrobehavior, namely, the formation of cities. The macrobehavior of cities emerging in these models appears to be insulated, at least to a certain degree, from the micromotives of the agents. This is a somewhat extreme example where the "details do not matter," and it shows how complex adaptive social systems may be subject to very strong aggregation forces that result in macrolevel emergence. It also demonstrates how such systems can embody deductive indeterminism, in the sense that a particular consequence (people wanting to live in cities) may have multiple, irreconcilable causes (they either like one another or hate one another). When the details do not matter, it is hard to deduce cause.

9.5 NETWORKS

In our earlier models of majority rule we implicitly introduced the concept of a *network*. In that simple, one-dimensional world the network linked each agent to k neighbors on each side. We saw how slight changes to the network (by altering the parameter k) altered the behavior of the model—increasing the size of the neighborhood decreased the number of blocks that formed, implying that the more connections across the society, the more cohesive it became.

The k-neighbor network explored in the Majority Rule model represents just one of many possible network configurations. Rather than forcing nearest-neighbors to be connected, we could have randomly connected the agents to one another. If we create these connections (or *edges* in the language of graph theory) by, say, allowing each possible

---

[6]A slight qualification here is that a *lone* agent is willing to remain where it is if it has an equal number of agents on either side of it. A city (more than one agent) cannot form at this site as the symmetry will be broken by the additional distance gained by leaving the other agents at the site.

connection to form with a fixed probability, we get an Erdos-Reyni network. Alternatively, we could make the connections according to some other distribution, such as a power law. In this case, the most connected person might have twice the number of connections as the second-most connected, three times as many as the third most, and so on. Networks as diverse as links to World Wide Web pages, metabolic pathways in cells, telephone calls, and sexual contacts all appear to have power-law-like distributions. Power-law networks have a fairly rich set of connections and are very resilient to random failures. However, such networks are susceptible to targeted failures: if you knock out the ten most connected nodes, you can significantly alter the system's behavior (Newman, 2003).

Another network structure that has been the focus of much recent attention is the *small-world* network. In a small-world network, each agent is first connected to a set of neighboring agents (as was the case in our Majority Rule model). To finalize the network, some of these local connections are randomly severed and replaced by random connections to anyone in the system. Thus, agents tend to be mostly, but not completely, locally connected.

If a network is solely composed of neighborhood connections, information must traverse a large number of connections to get from place to place. In a small-world network, however, information can be transmitted between any two nodes using, typically, only a small number of connections. In fact, just a small percentage of random, long-distance connections is required to induce such connectivity. This type of network behavior allows the generation of "six degrees of separation" type results (Watts, 2003), whereby any agent can connect to any other agent in the system via a path consisting of only a few intermediate nodes.

Next, we explore some simple network structures and examine their implications for two of the models we have already considered: the Majority Rule model and Schelling's Segregation model. We consider four different network structures:

- *Loop:* agents live on a circle and are connected to their immediate neighbors in each direction.
- *Grid:* agents live on a checkerboard-like grid where the edges of the grid wrap around to form a torus. An agent is connected to its immediate neighbors.
- *Pack:* agents exist in packs. Each agent is connected to its pack mates, as well as one agent outside of the pack.
- *2Loop:* agents live on two circles (both randomly ordered) and are connected to their nearest neighbors in each direction on both circles.

Figure 9.1. A segment of a Loop network. Agents live on a circle and, in this case, are connected to their two nearest neighbors on each side.

The Loop network is a common structure used in modeling. We have used it in our models of forest fires and segregation. In the Loop network, agents live on a circle and are connected to their nearest neighbors in each direction. A segment of the circle can be represented by a line as seen in figure 9.1. In real social worlds, a Loop structure might be appropriate for some (literal) neighborhood issues, such as the exterior color choices of the houses, the planting of gardens, lawn maintenance practices, and whether homeowners put up holiday lights. The Loop's regular structure greatly simplifies mathematical analysis; for example, in our earlier models of Majority Rule, these types of networks induced neighboring agents to take the same actions as one another, resulting in large, contiguous spatial chunks of identical behavior.

Glaeser, Sacerdote, and Schienkman (1996) used a Loop network and an assumption of majority rule to capture the spatial nature of criminal activity. In their model, each agent begins as either a criminal or a law abider. Initially, criminals are randomly distributed around the circle. Agents are then allowed to alter their behavior so that it agrees with the neighborhood majority.

It can be shown that even in networks with similar initial distributions of criminals, you can get very different final distributions of criminal behavior: some worlds become crime ridden while others become relatively crime free. This suggests that high crime rates in one area could be an artifact of unfortunate historical accidents rather than some difference in initial criminal behavior.

In figure 9.2 we illustrate two Loop networks that have the same number of criminals (action 1) but in different configurations. We assume that the agents look to their two nearest neighbors on each side and adopt the majority action. In the first network the society is slowly taken over by criminals, due to the driving forces induced by the criminals located at the left- and right-hand sides. In the second network the system maintains its initial state, stabilizing on the original number of criminals. As is clear from this example, to understand why one neighborhood becomes crime ridden while another does not, requires knowledge of both the initial predispositions of the agents as well as the structure of the influence network.

In the Grid network agents reside on a checkerboard. The top edge of the board is connected to the bottom edge, while the left side is connected

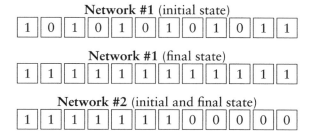

Figure 9.2. Dynamics of a Loop network. Each agent takes the majority action of the group formed by itself and one neighbor on each side. Here action 1 is becoming a criminal and action 0 is remaining law abiding. Agents update simultaneously each period.

to the right, causing the grid to take the form of a torus. An agent is connected to its immediate neighbors. The Grid is a simple extension of the Loop with agents being connected to their immediate neighbors (using two dimensions rather than one).

The Grid is a natural structure for activity that takes place on a flat geography (and it is also easy to represent on chalkboards and computer screens). Perhaps the most famous Grid model is the Game of Life (in this case the agents are influenced by the eight adjacent cells). Our earlier Standing Ovation model also relied on a Grid network.

In the Pack network agents live in tight cliques of mutual friends, as depicted in figure 9.3. Each pack is arrayed in a Loop network (thus, at each location on the circle is a pack rather than an individual). Within each pack, agents are connected to all of their pack mates. Agents are also connected to one other agent outside of their pack. To keep the arrangement regular, this outside connection is always to the corresponding member in the next pack ahead on the loop.

To formalize this network, consider a world with sn agents where s is the number of agents in each pack and n is the number of packs. We can number the packs from 1 to n and the agents within any given pack from 1 to s. Each member of pack is connected to all of its pack mates as well as its corresponding number in the next highest pack (or, in the case of pack n, the agent in pack 1). A Pack network has a small-worlds-like structure—each agent has a group of localized connections along with one connection outside of the local group.

Our final network, the 2Loop, consists of two distinct Loop networks. In both loops, agents are connected to nearest neighbors on each side. Agents are initially placed on each loop in random order. Thus, the nearest neighbors on the two loops are not necessarily related to one another.

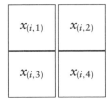

Figure 9.3. Part of a Pack network with four agents per pack. Each agent is connected to its three other pack mates, as well as the corresponding pack member in the next highest pack. Thus, agent $x_{i,j}$ is connected to all of the other agents in pack i as well as agent $x_{i+1,j}$.

The 2Loop network enables us to capture the effects of overlapping relationships in a network. For example, a teenage child may feel social influences from both her family and her friends.

9.5.1 Majority Rule and Network Structures

To explore how network structure influences system behavior we return to our Majority Rule model. In this model each agent must choose an action in $\{0, 1\}$. Agents switch their action if they find themselves in the minority. Here we assume that agents update their actions synchronously, though we again note that updating rules can influence the behavior of the model (see, for example, Huberman and Glance, 1993, and Page, 1997).

COMPUTATIONAL RESULTS

We first explore the model using computational experiments. We assume four hundred agents (the qualitative findings seem to be invariant across wide variations in this value). Recall that in each network, each agent has exactly four connections. Thus, any differences in outcomes are due to networks.

Define *conformity* as the percentage of agents taking action 1. Table 9.6 shows the average (across 1,000 trials) equilibrium conformity observed under the four different network types. As expected, initial conformity and equilibrium conformity are positively correlated. However, we see significant differences in conformity across the different network types. Given any initial level of conformity, we find that 2Loop networks tend to generate the highest equilibrium conformity followed by the Grid, Pack, and finally Loop network structures.

Features of the underlying network topologies can be used to explain the differences in conformity observed. In particular, note that the different network structures imply very different patterns of shared

TABLE 9.6

Equilibrium Conformity across Different Network Structures: Averages (Standard Deviations) across 1,000 Trials

Initial %	Grid		Loop		2Loop		Pack	
51	53.24	(0.23)	51.94	(0.27)	53.29	(0.20)	52.40	(0.28)
55	65.37	(0.25)	60.50	(0.26)	66.29	(0.19)	61.53	(0.27)
65	87.79	(0.29)	79.61	(0.15)	91.62	(0.15)	81.79	(0.18)
75	97.15	(0.31)	92.48	(0.06)	99.41	(0.09)	93.47	(0.04)

Note: Each of the 400 agents were randomly assigned action 1 based on the probability given in column 1. The system was then allowed to run until the conformity measure stabilized.

relationships. To make this concrete, we can count the intersection of an agent's connections (including itself) to the connections of its connections—call the average of this number the *overlap* of the network. For example, consider an agent in a Loop. A nearest neighbor has connections to four agents in common; the next nearest neighbor shares three connections. Thus, the overlap is equal to 3.5. In a Pack, the three neighbors in the same pack share four agents in common, and the one in the other pack has none; therefore, the overlap is 3.0. In a Grid, each neighbor shares only two common connections, leading to an overlap of 2.0. In a 2Loop, each neighbor typically shares two connections as well, resulting in an overlap of 2.0.[7]

The larger amount of overlap in the Loop and Pack networks provides a certain degree of insularity in the system. When the overlap is high, once the agents that overlap form a majority they will maintain it. Thus, once three or more consecutive agents in a loop take the same action, they will lock into that action regardless of outside forces. Similarly, three pack mates taking the identical action are insulated from the activities of the other packs. Thus, as the amount of overlap increases, we would expect more diversity in overall actions across the groups, and thereby lower conformity. That is exactly what table 9.6 shows. The Loop and Pack networks have the least conformity.

ANALYTIC RESULTS

In recent years, complex systems scholars have proved many mathematical results about the structure and function of networks. Here we illustrate how to derive some analytic results using two of the network structures we have described.

---

[7]Because the agents are randomly placed on the two loops, there is some probability that neighbors will be neighbors of each other, which would cause the 2Loop overlap to rise above 2.0.

Period 1:	M	0	1	M	1	1	M	M	M	1	M	0	M	1
Period 2:	0	0	1	1	1	1	M	M	1	1	0	0	1	1
Period 3:	0	0	1	1	1	1	M	1	1	1	0	0	1	1
Period 4:	0	0	1	1	1	1	1	1	1	1	0	0	1	1

Figure 9.4. Two-Pack dynamics. Each pack consists of two agents, each of whom is connected to the corresponding pack member in the pack ahead. In a given pack, agents are either both taking action 0, both taking action 1, or doing opposite actions (M). Given the dynamics of the model, once a pack does 0 or 1 it will lock into that behavior forever. Packs doing M always copy the current behavior of the pack ahead next period.

We begin by proving a result for our Pack model. Consider a version of the model where each pack has two members. As before, each agent is connected to its pack mates (here, the one other agent in the pack) as well as to the corresponding agent in the pack ahead. Note that the connection outside of the pack influences an agent's behavior only when the agent and its fellow pack mate take different actions. If the agent and its pack mate ever take the same action, the two of them will be locked in majority agreement thereafter.

Suppose that an agent is doing the opposite of its pack mate. In this case, it will always copy the action of its corresponding agent in the pack ahead.[8] Given that both members of the pack follow this behavior, this implies that whenever the two pack mates disagree on actions, then in the next period the pack will mirror the current behavior of the pack ahead of it in the loop.

Figure 9.4 illustrates these dynamics. Initially, agents take random actions. In the next period, packs that have both agents doing the identical action (designated in the figure as either 0 or 1 depending on the common action) remain unchanged; packs that are mixing their actions (M) copy the behavior of the pack ahead. Thus, as long as at least one pack has agents doing the identical action, the system will eventually converge to an equilibrium where each pack has its agents taking identical actions, though the chosen actions may vary from pack to pack.

---

[8]If the agent in the pack ahead is doing the opposite action, then the agent in question is in the minority and will want to change its action. If the agent in the pack ahead is doing the same action, then the agent in question will try to remain in the majority by not altering its action.

Given the dynamics, packs that end up with both agents taking action 1 do so via one of two possible routes. The first route is for both agents in the pack to have initially started out by taking action 1. The second route requires both agents to initially do opposite actions but eventually end up copying the pack ahead with both of its agents taking action 1. Using this reasoning, we can make the following claim for systems consisting of infinite numbers of packs:

Claim 9.5.1 *In the two-neighbor Pack model, the final conformity equals* $\frac{p^2}{1-2p+2p^2}$, *where* p *equals the initial probability that an agent takes action 1.*

To understand this claim consider the following. The initial probability that the two agents in a pack both take action 1 equals p^2 and the probability of them both taking action 0 is $(1 - p)^2$. Therefore, of all packs that have members doing the identical actions, $\frac{p^2}{1-2p+2p^2}$ of them have the agents coordinating on action 1. This probability also represents the chance that any pack that begins with its agents doing opposite actions will find the first coordinated pack to its right coordinating on action 1. Therefore, in equilibrium the proportion of packs coordinating on action 1 will be given by $\frac{p^2}{1-2p+2p^2}$. This proportion equals the conformity because all of the remaining packs are locked into action 0.

Conveniently, this same type of constructive proof can be extended to any Pack network with an even number of agents in each pack. If a majority of the agents in the pack takes the same action in one period, thereafter all agents in the pack will take that action. If the agents in the pack are split evenly between the two actions, then they all copy the respective actions of the pack members ahead. Thus, in a four-neighbor Pack model, final conformity equals $\frac{4p^3-3p^4}{1-6p^2+12p^3-6p^4}$.[9] This latter analytic result accords well with our computational results; for example, with $p = 0.65$, the theoretical conformity equals 81.65 percent while the computational experiments generated a value of 81.79 percent.

We can perform a similar analysis for the Loop network. Again, consider a two-neighbor version to keep the mathematics tractable. As before, assume that the initial probability that an agent takes action 1

---

[9]The initial probability that a majority of the pack will take action 1 equals $p^4 + 4p^3(1 - p) = 4p^3 - 3p^4$. The probability that a pack has a majority of its members taking action 0 equals $(1 - p)^4 + 4p(1 - p)^3 = 1 - 6p^2 + 8p^3 - 3p^4$. The result follows immediately.

equals p. In a two-neighbor Loop, any two neighboring agents that take the identical action will do so forever, as they form a majority with one another. Given this, we can derive the equilibrium conformity:

Claim 9.5.2 *In the two-neighbor Loop model, the final conformity equals*

$$\frac{2p^2 + p^3 - 3p^4 + p^5}{1 - p^2 + 2p^3 - p^4}$$

To prove this claim, consider the following. First, note that any agent that takes the same action twice in a row will do so forever, as it must be the case that at least one of its neighbors is doing the same action. Given this, any given agent can find itself in one of four possible states: it has just switched and taken action 1, it is locked into action 1 forever, it has just switched and taken action 0, or it is locked into action 0 forever. Second, note that any agent that has just switched to a different action will either lock into that action forever or switch to the alternative action in the next period. If an agent has just switched to action 1, then with probability $\rho_1 = 1 - (1 - p)^2 = 2p - p^2$ at least one of its neighbors will take action 1, locking the agent in question into doing action 1 forever; otherwise, with probability $\rho_2 = (1 - p)^2 = 1 - 2p + p^2$ it will switch to action 0. If an agent has just switched to action 0, it will either stay at zero forever, or with probability $\rho_3 = p^2$ it will switch back to action 1.

Let W_t and Z_t give the proportion of agents that have just switched to taking actions 1 and 0 respectively in period t. (Note that $W_0 = p$ and $Z_0 = 1 - p$.) Because ρ_1 is the probability that an agent that has just switched to doing action 1 will lock into doing it forever, $W_t\rho_1$ gives the proportion of agents that are newly locked into doing action 1 at time $t + 1$. For this claim, we want to know the asymptotic proportion of agents that have locked into doing action 1, which is given by $\sum_{t=0}^{\infty} W_t\rho_1$. For odd values of t, it can be shown that $W_t = Z_0\rho_2^{\frac{t-1}{2}} \rho_3^{\frac{t+1}{2}}$. For even values of t, $W_t = W_0\rho_2^{\frac{t}{2}} \rho_3^{\frac{t}{2}}$. Note that $\sum_{t_{odd}} W_t = Z_0\frac{\rho_3}{1-\rho_2\rho_3}$ and $\sum_{t_{even}} W_t = W_0\frac{1}{1-\rho_2\rho_3}$. Therefore, $\sum_{t=0}^{\infty} W_t\rho_1 = (Z_0\frac{\rho_3}{1-\rho_2\rho_3} + W_0\frac{1}{1-\rho_2\rho_3})\rho_1$, which, after substituting and simplifying, gives the required result.

Using these results, we can compare the expected conformity in the two-neighbor Loop and Pack networks:

Claim 9.5.3 *A two-neighbor Pack network leads to higher conformity than a two-neighbor Loop network.*

Based on the prior two claims, this claim holds if and only if

$$\frac{p^2}{1-2p+p^2} > \frac{2p^2 + p^3 - 3p^4 + p^5}{1 - p^2 + 2p^3 - p^4}.$$

Multiplying through and simplifying this inequality reduces it to

$$1 - 3p + 7p^3 - 7p^4 + 2p^5 < 0.$$

We can factor the left-hand side as $(1 - 2p)(1 - p)^2(1 + p - p^2)$. Because the first term is negative for $p > 0.5$ and the next two terms are positive for all p, the product of the three terms must be negative verifying the claim. This claim is consistent with what we found computationally for larger networks, namely that Pack networks lead to higher conformity than Loop networks.

9.5.2 Schelling's Segregation Model and Network Structures

We next analyze Schelling's Segregation model under alternative network structures. In the model there are two types of agents, A and B, who initially occupy random locations in the network. Exactly half of the sites are inhabited by residents. Each agent only cares about the percentage (of all occupied sites) of its neighbors that are of the same type. If this percentage falls below 40 percent, then the agent will randomly relocate to an unoccupied location.

Given the geographic nature of Schelling's model, we restrict our analysis to the Loop, Grid, and Pack networks (the 2Loop network does not naturally lend itself to a sensible geography). In each network agents are connected to eight others. For the Loop network, you can think of houses around a lake, along a shore, or surrounding a park, where each agent incorporates the types of its four nearest neighbors on each side into its decision to move. The Grid topology was used by Schelling in his original paper and, like Schelling, we allow the neighborhood to consist of all eight adjacent locations. Finally, the Pack network emulates a city that is composed of distinct neighborhoods. In such a world, even physically adjacent houses might be located in "different" neighborhoods. In the model, each pack ("city block") is composed of eight agents.

Recall that the main finding of the earlier Segregation model is that the system easily becomes segregated even though each agent is tolerant of the other type. Perhaps this result is due to some peculiarity of the Grid network used by Schelling? Surprisingly, we find that the Grid network understates the tendency toward segregation. Recall that *segregation*

TABLE 9.7
Segregation across Different Network Structures

	Grid	Loop	Pack
Segregation	75%	78%	80%

Note: Segregation is the percentage of like-type neighbors after equilibration, averaged across all of the agents in the system. Each network had 400 agents (the Grid was 20 by 20 and the Pack had 50 packs of size 8), and the values represent averages across 500 trials (all differences are statistically significant).

TABLE 9.8
Tipping across Different Network Structures

	Grid	Loop	Pack
Tipping	14%	20%	31%

Note: The tipping measures are averaged across 500 trials. The networks were identical to those described in table 9.7.

is the average percentage of an agent's same-type neighbors after the system equilibrates. As can be seen in table 9.7, both the Loop and Pack networks lead to higher measures of segregation in the system.

The Grid network also leads to the lowest amount of tipping across the three systems. Here *tipping* equals the percentage of agent relocations that were induced by any initial moves. Thus, if the initial moves result in a lot of subsequent movement, the tipping measure will approach 100 percent. In table 9.8 we see that tipping is positively correlated with the equilibrium levels of segregation observed in table 9.7.

To understand these results we can again use the idea of network overlap. Of the eight connections to an agent in a Pack network, seven share the same eight neighbors and one shares none, giving an overlap of 7.0. In the Loop network, two connections share eight neighbors, two share seven, two share six, and two share five neighbors, implying an overlap of 6.5. In the Grid network, four connections share six neighbors and four share four neighbors producing an overlap of 5.0.

As overlap increases there is more pressure to segregate, since like-types beget like-types given the greater connectivity among the set of agents. This also implies that we will see more tipping, as once a relocation forces one person to move, it is likely to cause the remaining neighbors to relocate.

In both the Majority Rule and Segregation models, we see that network structure has a large impact on the behavior of the system: different networks induced very different systemwide behavior. Although it is often difficult to characterize different network topologies succinctly, we found that our measure of overlap provided a coherent basis from which to unravel the link between networks and behavior in both of the models.

This analysis of network structure is by necessity rather brief. Nonetheless, it illustrates how network structure can have a big influence on a system's behavior. Moreover, it shows how certain inherent characteristics of networks, such as agent overlap, can provide insight across a variety of systems. Models that ignore networks, that is, that assume all activity takes place on the head of a pin, can easily suppress some of the most interesting aspects of the world around us. In a pinhead world, there is no segregation, and majority rule leads to complete conformity—outcomes that, while easy to derive, are of little use.

9.6 SELF-ORGANIZED CRITICALITY AND POWER LAWS

One of the hallmarks of complex systems is the aggregation of local actions into well-defined global patterns, such as cities, neighborhoods, and voting blocks. One such generic pattern, which characterizes many natural and artificial systems, is a distribution of activity characterized by a power law. A system is subject to a power law when $\text{Prob}[X = x] \sim x^{-k}$. If x is the number of occurrences of some event of a particular size, then a power law would imply that the likelihood of this event is proportional to the size of the event raised to the $-k$th power. Thus, if $k = 1$, events of size 100 are one-hundredth as likely as events of size 1. This implies that power-law-governed systems are characterized by many small events and a few, rare big ones.

Power-law-like behavior has been found in a variety of systems, including the use of words in texts (in English, a few words like "the" and "of" are used very frequently), the distribution of income in a society, the size of cities, and the magnitude of earthquakes and forest fires. An example of a social power-law distribution comes from Richardson's (1960) studies on war. Table 9.9 shows data on war casualties. At first glance, there does not seem to be much of a pattern to these data. If we rely on the exponential approximations shown in the parentheses, however, we see that as the number of deaths decreases by a factor of ten, the number of wars increase by a factor of three. (A fairly good prediction of the number of wars, W_i, with a given number of deaths, D_i, is given by the power law $W_i = 6517 D_i^{-1/2}$.)

Empirically, we often find examples that satisfy a power law with the exponent equal to -1. One such case is the relationship between city population, P_i, and rank, R_i. Roughly, this relationship is given by

TABLE 9.9
Richardson's (1960) Data on Deaths in Warfare, 1820–1945

Approximate Number of Deaths	Number of Wars
10,000,000 (= 10^7)	2 ($\approx 2 \times 3^0$)
1,000,000 (= 10^6)	5 ($\approx 2 \times 3^1$)
100,000 (= 10^5)	24 ($\approx 2 \times 3^2$)
10,000 (= 10^4)	63 ($\approx 2 \times 3^3$)
1,000 (= 10^3)	188 ($\approx 2 \times 3^4$)

$P_i = c R_i^{-1}$, where c is a constant. This equation implies that city size times rank is a constant (that is, $P_i R_i = c$ for all i), and thus the ith ranked city has $1/i$ of the population of the largest city.

Power-law distributions are just one member of a class of fat-tailed distributions. A distribution has fat tails if the probabilities of extreme events are "abnormally" high, where by abnormally we mean literally not like a normal distribution. If we assume that a distribution is normal when indeed it is fat-tailed, then we will grossly underestimate the potential for extreme events (and discount those that do happen as rare anomalies). Suppose that we observe a system, like an aircraft or power distribution network, that routinely experiences a few failures. If we believe that the failures are driven by a normal distribution (here, with low mean and variance), then we would expect that on most days we will have only a few failures and on rare occasions experience some bad days with lots of failures. On the other hand, if a fat-tailed distribution underlies the system, then days with very high numbers of failures are far more likely than we would expect. A fat-tailed view of social systems would imply that the outbreak of large-scale wars, the overthrow of a government, a "correction" in the stock market, and similar events are driven by forces that are quite different from what we usually assume.

A number of theories suggest why we might see power-law-like behavior. For example, Simon (1955) constructed a simple model to explain the city-size distribution. He assumed that the probability of a new resident joining a city is proportional to the city's size—a simple and plausible assumption. Given this assumption, it can be shown that a power-law distribution results.

An alternative approach to explaining power laws, developed by Bak and his colleagues (see Bak, 1996, for a general reference), focuses on the notion of *self-organized criticality*. The microfoundations of this model appear quite plausible in many social scenarios. Moreover, this approach is of particular interest to us because it is formulated around an agent-based spatial model.

The key driving force behind self-organized criticality is that microlevel agent behavior tends to cause the system to self-organize and converge to critical points at which small events can have big global impacts. Such critical points are familiar to anyone who has ever built a house of cards; while initially such constructions go quite smoothly, at some point the structure goes critical and even the slightest jiggle causes it to collapse. Similarly, in self-organized critical systems, the agents throughout the system tend to be poised in critical states where small disturbances can trigger large relaxation events that may encompass any number of agents. Consider a city of card houses, with each house spaced close enough to the others so that if it falls some of its cards will hit the neighboring houses. As we add a card to a random house, it either becomes a bit more unsteady or falls. Sometimes when a house falls, its neighbors are steady enough not to be disturbed, while at other (much rarer) times, all of the houses are a bit shaky and the collapse of a single house propagates across the entire city and everything comes tumbling down. Intuitively, one can imagine that the behavior of such a system (measured by the number of houses that fall in any given period) might correspond to a power law.

9.6.1 The Sand Pile Model

Bak's original thought experiment for self-organized critical phenomena involved a table that is randomly sprinkled with sand. As the sand accumulates, it forms a pile that eventually reaches its angle of repose. At this point the pile is critical, and additional grains start localized avalanches that often result in some grains falling off of the table. If we plot the distribution of the number of grains of sand that fall onto the floor each period, we find that we get a power law with an exponent of -1.[10]

To construct an even simpler model of self-organized criticality, we assume that agents are arrayed on a one-dimensional linear lattice. We assume that each location can hold at most $T - 1$ grains of sand, and when it receives its Tth grain, it topples and sends a single grain of sand to each of its neighbors within k sites.[11] Thus, if $T = 6$ and $k = 2$, upon receiving its sixth grain of sand the site will topple and send out four grains of sand, one each to the two immediate neighbors on the

[10]Physicists have conducted experiments using various particles (see, for example, Frette et al., 1996). Experiments with sand are not consistent with the predicted behavior, most likely due to confounds from shape and water content. Systems composed of certain types of rice do appear to generate the predicted properties.
[11]To ensure enough sand to go around, we need $T \geq 2k$.

TABLE 9.10
Self-Organized Criticality with $T = 6$ and $k = 2$

Action	Configuration	Ejected Particles
Initial configuration	4 3 2 4 5 5	0
Add a particle to site 5	4 3 2 4 6 5	0
Initial toppling (site 5)	4 3 3 5 2 6	1
Next toppling (site 6)	4 3 3 6 3 2	2
Final toppling (site 4)	4 4 4 2 4 3	0
Add a Particle to site 6	4 4 4 2 4 4	0
Final configuration	4 4 4 2 4 4	0
Initial configuration	4 3 2 4 5 5	0
Add a particle to site 6	4 3 2 4 5 6	0
Initial toppling (site 6)	4 3 2 5 6 2	2
Next toppling (site 5)	4 3 3 6 2 3	1
Final toppling (site 4)	4 4 4 2 3 4	0
Add a particle to site 5	4 4 4 2 4 4	0
Final configuration	4 4 4 2 4 4	0

right and the two on the left. Any sand that flows over the end points is assumed to be lost to the system.

During each time period a single grain of sand is added to a randomly chosen site. In table 9.10 we show some sample dynamics for this system. The upper and lower panels of the table differ in the order (but not the location) of the two particles we added to the system. Note that the final configuration and the total amount of sand ejected from the pile are identical in these two scenarios. This observation will hold in general for this model, as it is a member of a class of Sand Pile models that has the Abelian group property (Moore and Nilsson, 1999). This insensitivity to order implies that many updating methods, such as location-based asynchronous and synchronous approaches, become equivalent. Thus, in this model many of the agent-interaction details do not matter.

Even the one-dimensional model can generate some amazing patterns of toppling. Consider the case where $T = 4$ and $k = 1$ and every site has exactly three grains of sand. If we drop a grain of sand at the left edge of the lattice, one grain falls off of this edge and the other one topples the site on its right, which in turn topples the next site, and so on, and it is as if the sites were acting like a long series of dominoes falling in order. If we add a grain to the center of the lattice, however, we see waves of topplings, as shown in table 9.11. These topplings resemble the disturbances in a pool when a stone is dropped in the middle and the

TABLE 9.11
Self-Organized Criticality with $T = 4$ and $k = 1$

Action	Configuration
Initial configuration	3 3 3 3 3 3 3
Add a particle to site 4	3 3 3 4 3 3 3
Wave 1	3 3 4 2 4 3 3
Wave 2	3 4 2 4 2 4 3
Wave 3	4 2 4 2 4 2 4
Wave 4	2 4 2 4 2 4 2
Wave 5	3 2 4 2 4 2 3
Wave 6	3 3 2 4 2 3 3
Stability	3 3 3 2 3 3 3

resulting waves move outward until they bounce off of the pool's edges. Once stability is achieved in the system, we see that the sites resemble the initial configuration with the loss of only a single grain of sand in the middle position.[12]

One key question is how the two parameters of the model, T and k, influence its behavior. Increasing T has almost no effect on the system. This is not surprising, since T plays a role only in the early stages of the dynamics, and once a site has more than $T - 2k$ grains, it never falls below that amount. The more interesting area for investigation is the impact of k. To investigate this issue, and to gain some more insights into the fundamentals of the model, we next consider a very simple sand pile.

9.6.2 A Minimalist Sand Pile

Consider a one-dimensional sand pile with N sites, where each site can contain at most one grain of sand (therefore, $T = 2$). If a site acquires two grains of sand, it topples and sends one grain into each of its nearest neighbors (thus, $k = 1$). This minimalist sand pile not only displays many of the key behavioral features common in the more complicated self-organized critical systems, but it is also mathematically tractable.

To begin the analysis, note that the system will not experience any toppling as long as its sites have either zero or one grain of sand each. Second, consider the immediate impact of adding a grain of sand to a site. There are only a few possible cases. First, if the site is empty, it just

[12]The system actually discharges two grains—one off of each end—but we added a grain to get the waves going.

becomes filled with no other impact on the system. If the site is filled, it will topple, and its neighbors either become filled (if they were empty) or topple (if they were filled).[13] From these observations we can gain intuitions on larger-scale dynamics. For example, if we have a lattice of all ones with one zero, then if we add a grain to a site adjacent to the zero, we will see the zero "walk" toward the side of the lattice where the grain was added; thus, if we have 101111 and add a grain at site 3, we see 102111, 110211, 111021, 111102, and 111110.

Moore and Nilsson (1999) prove the following:

Claim 9.6.1 *If you have empty sites at j and k (with $j < k$) and filled sites in between these two, then if you drop a grain of sand at one of the intermediate sites, h (where $j < h < k$), then j and k will become filled and the site at $j + k - h$ will become empty.*

When applying this claim, we consider sites 0 and $N + 1$ (the sites just off of the edges of the lattice) to be empty. Thus, in the preceding example with the lattice 101111, sites 2 and 7 are empty, so when we add a grain to site 3, site 2 will fill and site 6 $(2 + 7 - 3)$ will empty.

Claim 9.6.1 has some important consequences for this system. First, as noted previously, whenever we add a grain to an empty site, it becomes filled and no other impacts occur. If we add a grain to a filled site, claim 9.6.1 applies (since the filled site must be bounded somewhere by empty sites), and the two empty sites will fill and one of the intermediate sites will empty. If the two empty sites bounding the filled sites are both on the interior of the lattice, then the lattice will gain one filled site. If only one bounding site is in the interior, then the lattice will maintain the same number of filled sites, and the interior boundary will move toward the associated end point by at least one site. If the two boundaries are not in the interior, that is, if the lattice is completely filled, then the system will be left with a single empty site. Given this behavior, the following claim can be made:

Claim 9.6.2 *The system will converge to a configuration with at least $N - 1$ sites filled.*

Claim 9.6.2 follows from the observation that if there are more than two empty interior sites, additional grains must either increase the number of filled sites by one or move the empty sites toward the edges.

---

[13] For the two sites on the end points of the line, we can assume that the "neighboring site," just off of the edge of the lattice, is perpetually empty.

TABLE 9.12
An Avalanche with $T = 2$ and $k = 1$

Configuration
1 1 1 1 1 1 1 0 1 1
1 1 2 1 1 1 1 0 1 1
1 2 0 2 1 1 1 0 1 1
2 0 2 0 2 1 1 0 1 1
0 2 0 2 0 2 1 0 1 1
1 0 2 0 2 0 2 0 1 1
1 1 0 2 0 2 0 1 1 1
1 1 1 0 2 0 1 1 1 1
1 1 1 1 0 1 1 1 1 1

Note: The initial grain falls on site 3.

Since once the system has all but one site filled, an additional grain either changes the position of the empty site or fills the entire array (if the grain lands directly on the empty site). In this latter case, the next grain will create a single empty site.

This claim leads to an important insight. In general, there are 2^N possible configurations of a binary lattice like the one in this model. However, claim 9.6.2 implies that after a while, the system will be in one of $N + 1$ configurations (the lattice can have an empty spot at one of N possible sites or it can be completely filled). For example, in the case of $N = 30$, we have taken a space of more than a billion possible configurations and reduced it to thirty-one.

9.6.3 Fat-Tailed Avalanches

We can use the preceding analysis to consider the distribution of avalanches. To begin this analysis, consider the behavior of our minimalist sand pile (with $N = 10$), shown in table 9.12. A grain of sand is dropped on site 3, and an avalanche begins. The initial toppling causes two waves to propagate outward. These waves will continue outward until they are reflected back by hitting an empty site (or, equivalently, an edge). The waves will stop once the two leading edges of the reflections meet.

The waves move one site during each iteration, so the two leading edges are either diverging or converging by two sites each iteration. The sites bounded between the leading edges will alternate between being empty and being ready to topple, and the system will stabilize once the two reflections meet. Because each wave moves either outward or inward by one site per period, and the total distance between the boundaries is seven, the waves must meet after seven iterations. Each wave will have covered seven sites in this time period (and since they both began at site 3, the waves will meet at site 5 and destroy each other).

There is an easy way to count the number of topples caused by the addition of the grain of sand. In the case shown in table 9.12 there were fifteen topples. Let M equal the number of contiguous filled sites bounded between the two empty sites. Let d give the distance between where the initial grain lands and the nearest edge[14] inclusive. (Thus, this distance is given by the number of sites between where the grain lands and the edge plus one.) As can be inferred from table 9.12, during the initial propagation of the wave, we will see $1, 2, \ldots, d$ sites topple, implying a total of $d(d+1)/2$ toppled sites. The same values, only in reverse, will happen at the end of the waves. Finally, in between these two end points, a period lasting $M - 2d$ iterations, we will have d topples each iteration. Summing these values we get:

Claim 9.6.3 *Dropping a grain in a block of M consecutive filled spots will result in a total of $d(M - d + 1)$ topples, where d is the number of sites in from the nearest edge where the initial grain landed.*

From claim 9.6.3 we can derive the exact distribution of avalanche sizes. From claim 9.6.2 we know that we will eventually get to a configuration with only one or no empty spots, so we can concentrate on these $N + 1$ configurations. Label these configurations $A_0, A_1, A_2, \ldots, A_N$, where A_i is the configuration with the ith site empty and A_0 is the configuration with no empty sites. Note that by using claim 9.6.1 it is easy to show that if we are in configuration A_i and we add a grain of sand to a random location, then we are equally likely to end up in any of the possible configurations.

For concreteness consider a lattice with $N = 10$. From any configuration, we can use claim 9.6.3 to calculate the number of topples as a function of where the grain lands. We do so in table 9.13 and figure 9.5, where each row is for a different landing spot. Since the system will spend an equal amount of time on average in each of the eleven configurations, we can compute the long-run frequency distribution just by counting the frequencies in the table. Table 9.14 gives this theoretical distribution and an estimated distribution from an experiment in which we dropped $100, 000$ grains of sand.

---

[14]It turns out that the formula we develop here is symmetric around the distance, so d can be the longer distance as well.

TABLE 9.13
Avalanche Size Given Landing Spot and Configuration

Landing Spot	A_1	A_2	A_3	A_4	A_5	A_6	A_7	A_8	A_9	A_{10}	A_0
10	9	8	7	6	5	4	3	2	1	0	10
9	16	14	12	10	8	6	4	2	0	9	18
8	21	18	15	12	9	6	3	0	8	16	24
7	24	20	16	12	8	4	0	7	14	21	28
6	25	20	15	10	5	0	6	12	18	24	30
5	24	18	12	6	0	5	10	15	20	25	30
4	21	14	7	0	4	8	12	16	20	24	28
3	16	8	0	3	6	9	12	15	18	21	24
2	9	0	2	4	6	8	10	12	14	16	18
1	0	1	2	3	4	5	6	7	8	9	10

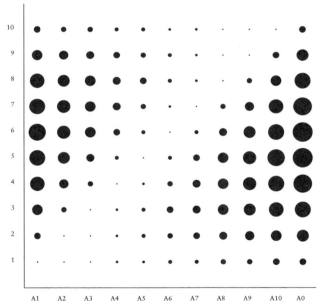

Figure 9.5. Graphical representation of avalanche size (larger circles represent larger avalanches), given the particle's landing spot (y-axis) and the system's initial configuration (x-axis).

TABLE 9.14
Theoretical and Experimental (100,000 Trials) Avalanche Distribution for $N = 10$

Avalanche Size	Theory (%)	Experiment (%)
1	2	2.0
2	4	3.9
3	4	3.9
4	6	6.1
5	4	3.9
6	8	7.9
7	4	4.0
8	8	8.0
9	6	6.0
10	6	6.0
11	0	0.0
12	8	8.1
13	0	0.0
14	4	4.0
15	4	4.0
16	6	6.0
17	0	0.0
18	6	6.0
19	0	0.0
20	4	4.1
21	4	4.0
22	0	0.0
23	0	0.0
24	6	6.0
25	2	2.0
26	0	0.0
27	0	0.0
28	2	2.0
29	0	0.0
30	2	2.1

The distribution in table 9.14 has fat tails, with the extreme events having a fair amount of weight. Also, from claim 9.6.3 we know that all avalanches are of size $d(M - d + 1)$, or in this case, $d(11 - d)$. This implies that certain avalanche sizes will not occur; for example, we will never have an avalanche greater than size 10 whose size is a prime number. If we increase the size, N, of the sand pile, then the number of times that we can have an avalanche of size R is roughly proportional to the number of divisors of R. If we plot this, we do not get a power law exactly, but the distribution does have fat tails.

The preceding analysis also gives some intuition about the time correlation we will see in avalanche size. The biggest avalanches we can get are of size 30. These occur when all sites are filled (A_0) and the grain is dropped in the center (either at site 5 or 6), putting the sand pile in configuration A_5 or A_6. In either of these states, the biggest avalanches that we can get are only of size 9. Thus, after a large avalanche we will see a small one. This type of correlation holds in larger models (including those with bigger k) as well. Large avalanches tend to unleash sites that are in critical or near-critical states and thus put the system in a state where only smaller avalanches are possible.

9.6.4 Purposive Agents

The previous model considers grains of sand dropping at random. What happens to the model with more thoughtful and directed agents? Suppose that agents have the ability to choose a landing spot and know the size of the resulting avalanche. Furthermore, assume that agents have preferences for either large or small avalanches.

Suppose the falling agents want to cause small avalanches. If the initial configuration is A_0 then the agent will want to land on either edge so as to generate an avalanche of size 10. The next agent will choose to land on the empty site created by the first one and will avoid causing an avalanche. The third agent is in the same situation as the first. Thus, we will find the system alternating between avalanches of size 10 and 0, for an average avalanche of size 5.

Alternatively, suppose the agents want to cause large avalanches. If the system is in A_0, the first agent will land at either site 5 or 6 and cause an avalanche of size 30. Suppose its choice leaves the system in A_5. In this case the next agent will want to land at site 8 and cause an avalanche of size 9. The system is now at A_8 ($5 + 11 - 8 = 8$), so the next agent will choose site 4, causing an avalanche of size 16 and putting the system in configuration A_4 ($0 + 8 - 4 = 4$). From here site 7 or 8 is the best choice, producing an avalanche of size 12. If we pick site 7, we return the system to A_8 ($4 + 11 - 7 = 8$).[15] Thus, we lock into a cycle of length 2 that alternates between A_8 and A_4 with associated avalanches of size 16 and 12.

Therefore, if agents attempt to maximize the the size of each avalanche, they fall into a two-cycle pattern with an average avalanche size of 14. However, this assumes that the agents play the game myopically, as better alternatives exist if they can break out of the cycle. A socially better cycle

---

[15]Picking site 8 results in two potential cycles with either the same or slightly lower payoffs.

begins in state A_0, followed by an agent landing on site 5 (or 6), resulting in an avalanche of size 30, and then having the next agent refilling the resulting hole and putting the system back in A_0. This results in a cycle with an average avalanche size of 15.

As these thought experiments show, with purposive agents the sand pile becomes malleable. Indeed, it is possible for agents to generate almost any distribution of avalanche sizes provided that they have enough foresight (Ishii, Page, and Wang, 1999).

9.6.5 The Forest Fire Model Redux

There are many other ways to form models that exhibit self-organizing criticality and fat tails. Consider the Forest Fire model discussed in chapter 7. Recall that each site on a lattice either has a tree or is barren. During each time period, barren sites grow a tree with probability g, and sites that already have a tree in place remain unchanged. After any new trees have grown, the fire season begins and there is an independent probability of f that a flash of lightning will strike each site and ignite its tree. Once a tree is ignited, it and any immediately adjacent trees burn until the site becomes barren. Burning trees, in turn, ignite their neighbors, and the fire will only be contained when it hits a barren site (literally hitting a fire wall).

This model tends to a critical state where fires of all sizes are exhibited, with the majority of fires being fairly small. On occasion, when there have not been many fires for a while, a large conflagration that encompasses most of the forest will occur.

Standard approximations of this problem that rely on the "expectations" of growth and fire are not able to capture the underlying statistics very well. The problem is that these approximations do not adequately account for the self- organization in the system. That is, lightning creates coherent clusters of either barren land or trees that have an underlying structure that is fundamentally different from the one approximated by independent random events.

In the model, the randomness in the microlevel processes makes possible coherent macrolevel patterns. These macrolevel patterns, while being driven by randomness, are not random in the traditional sense. By analogy, a rock is composed of atoms that are constantly shifting due to random events; nonetheless, it tends to be a very coherent object. If, by chance, each of the atoms on one side of the rock randomly shifted in the same direction at the same time, that part of the rock would instantaneously bulge outward. Notwithstanding the potential of such an event, we can view the rock as a coherent, nonrandom structure, created by a very incoherent randomness.

9.6.6 Criticality in Social Systems

The preceding models demonstrate how power-law-like behavior can emerge from systems of interacting agents. The generative mechanisms underlying such models seem appropriate for certain types of physical systems, like forest fires, earthquakes, mud slides, and floods, and perhaps they could also be justified for some social phenomena, like riots, traffic jams, and utility blackouts.[16] That being said, other examples of power-law-like behavior, such as city-size distributions and word counts in texts, probably come from very different generative mechanisms and thus may not be so easily explained by self-organized criticality.

Also, recall that adaptation can alter the critical behavior of a system. We saw how purposive agents can completely change the dynamics of the sand pile. We also found in chapter 7 that agents adapting their growth rates in the Forest Fire model significantly change the system's behavior. A closer examination of this latter system indicated that the locations tend to self-organize based on risk levels, with each location becoming either an absolute risk taker or avoider. With adaptive agents, the system configures itself in a way that mitigates the overall risk by preventing criticality from emerging. In essence, the adaptive actions of the individual agents lead the system away from the critical regime and more toward what an omniscient designer attempting to balance risk and stability would create.

In the Forest Fire model, the ability to adapt away from criticality was facilitated by the stable spatial interactions among the agents. The presence of risky neighbors forces an agent to avoid risk and, in the process, become an inadvertent fire wall. If instead we were to randomly scramble the locations each day, this coherence would be lost and we would see a very different outcome. Some social systems have much more stable interactions than others. Thus, we might expect that banks and suburban neighborhoods will be able to develop adaptively the necessary fire walls to prevent criticality, while the more transient relationships inherent in, say, urban neighborhoods and highway travel will not.

The models explored in this and the previous chapter had a number of key features. First, they crossed a broad swath of the complex systems research agenda and embraced many of the key ideas that have emerged in this area over the past decade or so. Second, they were presented in a very simplified form, both to keep them accessible and to allow one to make easy connections across the various domains. Finally, they illustrated a variety of methodologies and concepts that lend insight into how one can think about social science as a complex system.

---

[16] The theory offers little guidance as to why Los Angeles would embrace all of these events.

CHAPTER 10

Evolving Automata

> There is grandeur in this view of life, with its several powers,
> having been originally breathed into a few forms or into one;
> and that, whilst this planet has gone cycling on according to
> the fixed law of gravity, from so simple a beginning endless
> forms most beautiful and most wonderful have been, and are
> being, evolved.
> —*Charles Darwin, Origin of the Species*

> Mistakes are the portals of discovery.
> —*James Joyce, Dubliners*

MODELS WHERE AGENTS adapt their behavior based on experience are
very useful in the exploration of complex adaptive social systems. In
many social systems, agents are not static behavioral drones; rather,
they alter their behavior based on past feedback or the anticipation of
future events. A key scientific question is how does adaptation alter
the dynamics of complex systems. From a modeling perspective, the
introduction of adaptive agents provides a means by which to create
models that can explore new realms of agent behavior that transcend the
usual bounds imposed by the modeler. From a practical point of view, if
we can develop systems that are able to adapt effectively in a complex
social environment, then we can explore, design, and refine new complex
social systems, such as computer networks and auction mechanisms.

The approach we pursue here evolves agents controlled by simple
computer programs (Miller, 1988) using a genetic algorithm (Holland,
1975). Models based on artificial adaptive agents (Holland and Miller,
1991) have proved to be useful in a variety of social science fields,
including anthropology, economics, organizations, political science, and
sociology.

10.1 AGENT BEHAVIOR

The behavioral substrate for our agents is based on simple computer
programs modeled by finite automata. Finite automata are mathematical
models of systems that have discrete inputs and outputs; such systems

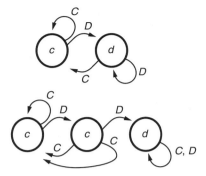

Figure 10.1. Two sample automata. States are given by the large circles, transitions by the labeled arcs, and actions are shown in the interior of each state. We assume machines start in the left-most state, and that states are labeled starting at 1 from left to right.

capture a fundamental class of behavior rich in applications and possibilities. Automata are sufficiently powerful that they can solve any problem that can be solved by a computer.[1]

The specific type of finite automata we use are Moore machines, though given the fundamental nature of these models, Moore machines can be easily transformed into various alternatives. Formally, a Moore machine is composed of a set of states S. Each state of the machine, $s \in S$, is associated with an output (agent action) that is produced each time the machine enters that particular state. Let λ define this association: $\lambda : S \to A$, where A is the set of allowable outputs. Also associated with each state is a transition function, δ, that determines the next state that the machine will enter based on the observed input: $\delta : S \times A_\sim \to S$, where $A_\sim$ gives the observed input. Finally, let each machine begin in a designated starting state (or initial state) prior to receiving any input.

A more intuitive description of a Moore machine is given by its transition diagram. Figure 10.1 illustrates two sample automata. The states of the automata are represented by the bold circles (and, for convenience, we number them consecutively from left to right starting at 1). The label inside each state gives the output that the automaton produces when it enters that state. The labeled arcs show the transitions among the states, with each label specifying a potential observed input. We assume here that each machine starts out in the left-most state

[1] It can be proved formally that not all functions can be solved by a standard computer (that is, not all functions are "computable"). We leave it as an open question whether any of the functions that are of most interest to complex adaptive social systems, such as the behavior of a typical social agent, are in this latter group.

(state 1). Thus, the upper two-state automaton begins by producing output c and continues that output as long as it receives input C. If the input changes to D, however, the automaton enters state 2 and produces d, and it will continue to produce this output as long as the input continues to be D. This machine is a Tit-For-Tat machine that begins by producing c and then subsequently mimics its opponent.

The three-state machine in the lower panel of the figure begins by producing c and will continue with this output as long as the input does not have two D's in a row. If two consecutive D's are observed, this machine enters state 3 and takes action d regardless of any future inputs. This machine implements a (somewhat tolerant) grim trigger strategy: begin with c but, if the opponent ever persistently plays D, do d for the remainder of the game.

Note that these finite automata do not have any stochastic components. Since they can emulate any computation, it is possible, though messy, to implement a pseudo-random-number generator within an automaton. Such "random" behavior is embedded in the machine through the addition of states. In practice, however, it is far easier to allow the machine access to an external random-number device if needed. Access to external random-number generators simplifies machine analysis, as it is often difficult to disentangle states used for randomization versus those of a more systematic nature. This problem becomes acute when we rely, as is commonly done, on the number of states in a machine as a proxy for complexity. While randomness is "complex" in some sense, it is important not to confound a hundred-state machine that embeds a sophisticated algorithm with one that is just trying to emulate a coin flip.

10.2 ADAPTATION

There are various ways to incorporate adaptation in our models. For example, agents could have prior beliefs over potential behavioral heuristics and update these beliefs as they experience payoffs, agents could use nonlinear algorithms to recognize key opportunities for action, and so on. Our goal here is not to provide a complete list of all of the ways in which agents might respond and predict, but rather to provide some examples of how to construct adaptive agents. Therefore, we focus on using analogs to evolution in natural worlds as a way to drive adaptation.

Evolutionary processes are relatively simple to implement and put few demands on an agent's abilities. Of course, the adaptive behavior of agents in real systems may be driven by much more sophisticated mechanisms than simple evolution. Nonetheless, the mechanisms explored here

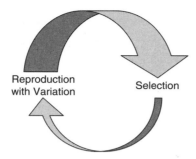

Figure 10.2. The "theory of evolution." During each generation, individuals in the population undergo selection based on a measure of fitness. These fitter individuals are then reproduced (amplified) with variation (perhaps from mutation and other genetic operators like crossover), and the generation begins anew.

represent a useful lower bound on adaptive behavior and serve as a convenient way to model adaptation.[2]

Here we will rely on a very simple form of adaptation modeled using a genetic algorithm (Holland, 1975; Mitchell, 1997, provides a useful introduction). Genetic algorithms model adaptive systems that are driven by variation and selection, analogous to the processes outline for the natural world by Darwin's theory (see figure 10.2). As in natural adaptive systems, genetic algorithms rely on a population of solutions that are replicated with random variations based on their performance (see figure 10.3).

Genetic algorithms belong to a class of population-based search algorithms. They do not have a single solution that keeps improving; rather, they rely on a pool of potential solutions. To employ a genetic algorithm, we must first develop a useful representation of the potential solutions. There are usually a variety of ways to represent such solutions. Good representations are those that admit a broad class of potential solutions, so as not to unduly limit the search. Moreover, good representations must be compact, consistent, and coherent. Compactness implies that with relatively few manipulations a large amount of the solution space can be explored. Consistency requires that simple variants of a given solution are also admissible (though, perhaps not high-quality)

---

[2]Evolutionary explanations for phenomena can be taken too far. There is a Darwinian alternative to the theory of gravity as an explanation for Newton's falling apple that floats around biology conferences: originally, apple trees used to shoot out apples in all possible directions, some heading for the sky, others running parallel to the earth, and so forth, and it was only those apples that fell downward that were able to grow and reproduce.

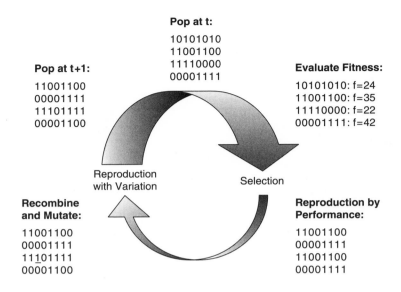

Pop at t:
10101010
11001100
11110000
00001111

Evaluate Fitness:
10101010: f=24
11001100: f=35
11110000: f=22
00001111: f=42

Pop at t+1:
11001100
00001111
11101111
00001100

Reproduction with Variation

Selection

Recombine and Mutate:
11001100
00001111
11101111
00001100

Reproduction by Performance:
11001100
00001111
11001100
00001111

Figure 10.3. A genetic algorithm. Each solution is represented by a bit string. Moving clockwise from the top, each potential solution in the population is tested on the problem and given a fitness value, solutions with greater fitness are selected for reproduction, and then they are either retained as is or modified by recombining parts of the selected solutions (crossover) and by making small changes (mutation). This completes one generation of the algorithm, which is then iterated based on the newly formed population.

solutions. Finally, coherence requires the solution space to be structured in such a way that variations of solutions are connected to one another in a meaningful way. That is, a coherent representation will ensure that "neighboring" solutions tend to share key structures with one another.

For example, consider the problem of creating a drawing of a person (see figure 10.4). We can represent potential solutions to this problem by a sequence of six components (for example, hair, eyes, nose, cheeks, mouth, and chin), each of which has ten different variations. Although this representation is relatively compact—being composed of only six features and a total of sixty separate pieces—it can represent one million (10^6) unique faces. It is also consistent in the sense that any set of feature choices will result in a recognizable face. Finally, there is a coherence to this representation: if we alter a single choice, say, switch the mouth, we get a face that looks related to the original.

It is equally helpful to consider a bad representation of the above problem. Suppose that we take all 10^6 faces and randomly assign each

Figure 10.4. Representing a face. Suppose that to draw a face we need to pick a chin, mouth, nose, ears, eyes, and some hair. If, for any of these elements, there are ten possible choices, we have only sixty pieces with which to form faces. Nonetheless, given the combinatorics, we can create 1 million different faces from these sixty pieces.

of them a number from one to 10^6. Going from, say, face number 27,182 to face 27,183 is a random move in "face" space, and thus lacks any coherence. Under this arrangement, neighboring faces look quite different from one another, and there is little structure for a learning algorithm to exploit.

Obviously, the efficacy of search is closely tied to the structure of the search space. As an analogy, consider searching for a hidden number between 1 and 10^6. If the space is well ordered—for example, each search reveals whether the hidden number is above or below the value searched—then with twenty, carefully selected searches, you can find the number. If, on the other hand, the space has no order, then on average it will take around 500,000 searches before the hidden number is found.

Given a representation, a genetic algorithm begins by randomly creating a population of possible solutions. The algorithm then proceeds through a series of generations. During a generation, each possible solution is given a performance measure on the problem. In the original work on genetic algorithms, this performance measure came from evaluating the solution on an exogenous objective function. In many social systems, however, performance is often endogenous as it depends on the actions of the other agents in the system. When we have endogenous performance, we are in a coevolutionary world, as changes in the behavior of one agent alter the environment and performance of all of the other agents.

Once each solution has been evaluated, the genetic algorithm selects solutions for reproduction. The key requirement for a selection mechanism is that better-performing solutions have a higher chance of being allowed into the next generation. Many mechanisms meet this criterion. For example, under "roulette" selection, solutions are given lottery tickets based on performance and then randomly selected for reproduction. Under "tournament" selection, two (or more) solutions are randomly chosen with replacement, and the best of these are reproduced. Selection mechanisms only need to be biased toward performance, and thus there is no guarantee that the best-performing solution will be reproduced or that the worst one will be eliminated. Of course, we could implement an algorithm in which the best solution always survives. This type of requirement may be more likely in social versus natural worlds, as social agents may be able to emulate the top solutions easily and rapidly, whereas in biological systems even the best solution can die without progeny.

Most genetic algorithms reproduce just enough solutions to keep the population size constant across generations. In any one generation, we will typically see multiple copies of the better-performing solutions in the population. Indeed, in a world with exogenous fitness and only the selection operator, the population will converge on a homogeneous agglomeration of the best-performing strategy that was present in the initial population.

Once a new population has been selected, the final step in a genetic algorithm is to introduce variation, using genetic operators, into some of the reproduced solutions. Two operators are commonly used: mutation and crossover. The mutation operator makes small random changes within a given solution. The crossover operator exchanges larger parts of two solutions. In the previous example of drawing faces, a mutation operator might randomly pick one of the six facial features and randomly replace it with one of its nine variants. Crossover, might swap the hair, eyes, and nose choices of one solution with their counterparts from another solution.

Crossover is often characterized as a source of random variation, but this is not entirely accurate. Although the crossover point (or points in some implementations) is random, the constituents of the pieces being crossed are not. Each of these pieces represents partial solutions that have survived rounds of selection, and as such these partial solutions represent better than average "guesses." For example, faces that have made it through many rounds of selection must have individual parts and combinations of parts that are better than average solutions to the problem at hand. Thus, while the switching of pieces is in some sense random, what is switched is not.

The actual mechanisms underlying genetic operators are an area of active investigation, and a clear understanding of their workings is slowly emerging. In general, mutation is viewed as a way to keep adaptive systems from getting trapped in narrow regions of the search space versus being a constructive way to find good solutions to problems. Thus, although most mutations are likely to be deleterious to performance (consider trying to improve your car's performance by randomly hitting the engine with a few hammer blows), mutation is nonetheless necessary, as on occasion it results in new opportunities (the hammer blow destroys the water pump and the air-cooled engine is born). The benefits of crossover stem from the fact that good partial solutions, known as building blocks or schema, are present in the population. The current theory of genetic algorithms suggests that crossover effectively preserves and combines building blocks, allowing good solutions to be built from the bottom up.

Once a new population of solutions is formed, the generation is concluded. The system is then iterated and a new generation begins, during which the cycle of evaluating performance, selection, and modification continues anew.

As should be apparent from this description, genetic algorithms are a broad class of algorithms, and many variations are possible. Various representations, selection mechanisms, and genetic operators are in common use, and researchers are continually working on new variations of key parts of the algorithm. Notwithstanding these variants, genetic algorithms appear to be fairly robust to any particular choice. Indeed, this robustness suggests that there may be large equivalence classes of adaptive behavior where the details do not matter.

Most computational models can rely on simple variants of the genetic algorithm and still meet modeling imperatives. For example, the simplicity of tournament selection and its ordinal use of performance make it a natural choice for the selection operator. The one choice that appears to have the biggest impact on model behavior is how the potential solutions are represented—good representations give the genetic algorithm a better chance to evolve quality structures. An interesting side effect of coming up with a good representation is how such structures can form the basis for thinking about the problem more generally. For example, a good representation might suggest a new way to apply mathematical tools to the problem or more general connections to other interesting problem domains.

10.3 A Taxonomy of 2 × 2 Games

For our first application of evolving automata, we consider strategic play in simple, repeated games. The formal analysis of strategic play in games has been around for a long time, dating at least back to fifteenth-century Japan with the establishment of government-subsidized

TABLE 10.1
A Sample Payoff matrix (Prisoner's Dilemma)

		Column Agent	
		A_c	B_c
Row	A_r	2,2	4,1
Agent	B_r	1,4	3,3

Note: Payoffs are ordered row agent, column agent.

Go schools. Game theory began to enter the main stream of many fields with the development of mathematical techniques prompted by World War II. Recently, game theory has become a central theoretical tool in fields ranging from biology to economics.

One important class of games is composed of repeated, two-player, two-action (repeated 2×2) games. This very simple class serves as the foundation for key parts of both theoretical and applied game theory. Despite much effort, though, our understanding of strategic behavior in these games is unfortunately very limited. Theoretically, mathematical results like the Folk theorem suggest that the behavior of rational agents is relatively unconstrained. Experimentally, only a few games from this class, such as the Prisoner's Dilemma and some coordination games, have been widely studied. Given both the importance of repeated 2×2 games and the limitations of current theoretical and experimental tools, we need to find new avenues for exploring this area. Here we advocate the application of computational models of evolving automata as a valuable means for conducting such investigations.

In a basic (nonrepeated) 2×2 game, the two players each have a choice between two possible actions and must choose an action without any knowledge of what the other agent will pick. Once the agents have made their choices, each receives a payoff that depends on the joint actions. A convenient way to summarize this scenario is by arraying the payoffs in a matrix with the choice of row and column being assigned to the actions of the "row" and "column" agent. Table 10.1 shows a sample payoff matrix where, for example, if the row agent chooses action A_r and the column agent picks B_c, then the agents will find themselves in the upper-right corner of the matrix. In this corner, the row agent receives a payoff of 4 and the column agent receives a payoff of 1.

The payoffs shown in Table 10.1 define the well-known Prisoner's Dilemma game. In this game, each player always earns a higher payoff by choosing action A regardless of what the other player chooses. If both agents make this choice, however, they each receive a payoff of 2 (upper-left corner). If they had both picked action B, they each would have earned a payoff of 3 (lower-right corner).

In a repeated 2×2 game the agents play the same 2×2 game over and over. In each round of play the players simultaneously choose actions. The actions are then revealed, each agent accumulates the associated payoffs, and a new round is begun.

Our goal here is to use evolving automata to analyze strategic behavior in repeated 2×2 games. Rather than investigating a specific game, such as the Prisoner's Dilemma (see Miller, 1988, 1996 for such an analysis), we use the power of computational experiments to investigate strategic behavior across an entire class of games. Since payoffs are associated with real values, there are an infinite number of 2×2 games, and we must therefore find some reasonable way to pare down these games to make this investigation feasible.

Rapoport and Guyer (1966) developed a taxonomy of 2×2 games by assuming that agents have only ordinal preferences over outcomes. Under this constraint there are seventy-eight unique games.[3] Twelve of these games have symmetric payoff matrices. The taxonomy includes analogs to almost all of the most widely studied games, though it does exclude a few important ones, such as Matching Pennies. Also, the resulting set may not be a fair and balanced selection across all games as some game types are overrepresented in the final taxonomy.

For repeated games, ordinal preferences over one-period outcomes do not sufficiently specify preferences. Thus, for the analysis here we transform payoffs into cardinal measures drawn from the set $\{1, 2, 3, 4\}$. By imposing a specific set of cardinal values we may impede our ability to generalize the results. However, the algorithm we use depends only on the cross-agent rankings of the final payoffs, so it can withstand some types of transformations of the payoff matrix. Moreover, our goal here is to understand general patterns across the class of games, and many of these patterns are likely to be robust to many cardinality assumptions.

10.3.1 Methodology

To analyze strategic behavior across the seventy-eight games we use a computational model of evolving automata based on the methodology developed by Miller (1988, 1996). In this methodology, each

[3] Ordinal preferences imply that there are 24 (4!) possible orderings in the payoff matrix for each agent. Thus there are 576 (4! × 4!) possible games. These 576 games reduce to 78 once various symmetries are recognized.

agent's game strategy is based on a sixteen-state Moore machine (see section 10.1) where the output of the automaton gives the agent's desired action in the round and the input comes from the opponent's last action. Agents are matched in round-robin tournaments against every other agent and accumulate payoffs from all of their games. At the end of the tournament, the entire population of agents is modified using a genetic algorithm (see section 10.2).

For each of the seventy-eight games we randomly create two populations of size 50, one for the row and one for the column players. Every row player plays every column player once in a fifty-round[4] repeated game. An agent's final payoff in the tournament is given by the sum of all of the payoffs it receives during each of its fifty separate matches.

At the end of the tournament, the two populations of row and column players are separately modified by a genetic algorithm. Agents are selected for modification based on their total payoffs. Two agents are randomly chosen (with replacement) and the one with the higher score is allowed to go into the next generation. This selection procedure is iterated fifty times, so that a new population of fifty agents is formed. Each agent in the new population has a 50 percent probability of being modified. An agent is modified by taking a random state of its machine and with a 50 percent probability altering the action in this state. The remainder of the time, one of the state's two transitions is randomly selected and redirected to a random state. We do not use any kind of crossover here, as the space of strategies is sufficiently small that mutation alone can drive the search. Once each agent has had a chance of modification, the generation is completed. This procedure is iterated for 150 generations. For each of the seventy-eight games we conduct fifty separate experiments, as described previously. In the analysis presented here, the final data are means over the fifty trials.

The number of states used in a machine has been suggested in the literature as a measure of strategic complexity. Alas, this measure is rather indirect for a few reasons. First, isomorphic machines can have different numbers of states. For example, a sixteen-state automaton with all states generating a C output is equivalent to a one-state automaton with a C output (and transitions back to this single state). We can circumvent this problem by relying on a theorem from the study of automata that shows how any machine can be represented by an isomorphic minimal-state machine. A second problem with using states to proxy for complexity is that the transitions among the states also matter. For example, a machine that plays C nine times in a row and then

---

[4]The size of the Moore machines makes it impossible for agents to recognize the finiteness of the repeated game.

plays D thereafter requires ten states (the first nine states output C and transition to the next one in the sequence regardless of input, and the last state outputs D and transitions to itself). Similarly, a machine that plays C ninety-nine times in a row and then D thereafter requires one hundred states. Even though the number of states between these two strategies differs by a factor of ten, intuition suggests that their underlying strategic complexities are quite similar.

While the focus of the analysis is on the general behavior across all of the games, considering the behavior that arises in a specific game like the Prisoner's Dilemma is instructive. Initially, the play of the game reflects the random nature of the strategies. Relatively quickly, agents learn to defect in the game and the system is characterized by high levels of mutual defection. However, cooperative strategies soon begin to emerge and we see a transformation of the system toward mutual cooperation. An in-depth analysis of this phenomenon indicates that Tit-For-Tat-like strategies (see top of figure 10.1) emerge. Such strategies can survive in a world of defectors, as they quickly fall into mutual defection when facing such opponents. When they meet another Tit-For-Tat-like strategy, however, they achieve mutual cooperation and do quite well.

10.3.2 Results

Our analysis using evolving agents suggests that the majority of the games in the taxonomy have very predictable outcomes. Of the seventy-eight games, in 59 percent agents concentrated more than 99 percent of their moves on a single outcome by the end of the evolution. A review of these games suggests that not only are they "boring" to our agents, but as it turns out they have not been all that interesting to social scientists as well—very few research experiments have been conducted on these games.

As the agents evolve we may see large changes in how they play the game, what we call outcome dynamics. Large variations in the outcome dynamics are indicative of games in which adaptive behavior yields changing strategic regimes over time. Approximately 25 percent (19/78) of the games displayed above-average variation in outcome dynamics. The Prisoner's Dilemma clearly had the highest variation of all of the games (with a measure almost two standard deviations above the next highest). The four games with the highest variations consisted of all the games with a single, iterated dominant strategy equilibrium that was Pareto dominated by a non-Nash outcome. Besides the Prisoner's Dilemma, two other symmetric games had above-average measures of outcome variation: Chicken, which was fifth, and Battle of the Sexes, which was sixteenth.

Our analysis identified three (out of the twelve) symmetric games as having intriguing outcome dynamics: Prisoner's Dilemma, Chicken, and Battle of the Sexes. These are also the three games that have dominated the experimental and theoretical literature over the past decade or so. While each of these games is celebrated because of its "natural" applications, the evolutionary dynamics explored here suggest that they are also naturally interesting.

In some games, small changes had definite impacts on the evolution of play. For example, two of the games are symmetric coordination games with the two coordination points yielding (4,4) and (3,3). In one of these games, if the row player attempts to play for the (4,4) outcome and the column player goes for the (3,3) one, the payoffs are (1,2), while in the other game the corresponding payoffs are (2,1). Thus, in the latter game it is a little less costly to try for the (4,4) outcome and not coordinate. This small change results in 99.3 percent of the plays being concentrated on the (4,4) outcome in the latter game, but only 94.3 percent in the former.

Different patterns of overall strategic complexity (measured by minimized automaton size) were noticeable across the games. Games with the lowest levels of strategic complexity were those with dominant strategy equilibria that were not Pareto dominated. Games with the highest complexity were those without pure strategy equilibria. This latter group was one of the few times that Rapoport and Guyer's classification scheme closely corresponded to a particular adaptive behavior. There was some positive correlation ($\rho = .35$) between games with high outcome variation and those with high strategic complexity.

This analysis indicates that interesting patterns of adaptive strategic behavior can be identified in the class of repeated 2×2 games. We find that about 25 percent of these games yield game dynamics that exhibit large changes during the course of adaptation. It appears that the Prisoner's Dilemma and its three asymmetric variations have the most volatile dynamics. The analysis of strategic complexity indicates that certain types of games may be strategically easier than others to play. Those games that do exhibit a lot of strategic adaptation during evolution, however, do not necessarily end up with the most complex strategies.

Heretofore, the analysis of even simple repeated games has not resulted in a broad understanding of strategic behavior. Theoretically, few predictions emerge; experimentally, few games can be studied. Obviously, new routes of analysis must be explored. One such avenue relies on the study of artificial adaptive agents. The analysis of such systems permits us to derive and generalize key patterns of strategic behavior embodied by simple adaptive learning systems.

10.4 GAMES THEORY: ONE AGENT, MANY GAMES

In game theory the tendency has been to focus on agents playing one game at a time. Using evolving automata, we can easily model "cognitive" behavior across multiple games. Imagine that agents must play many games, not just one; thus, at one moment they play, say, a Prisoner's Dilemma and at the next a game of Matching Pennies. The important point here is not the precise composition of the games but rather the requirement that a single agent faces multiple games.

Think of an agent's Moore machine as a "brain" that must be allocated across, say, two games. This could be accomplished either by the agent compartmentalizing different sections of its machine for each game or by it somehow sharing the same parts across the games. If it must share the parts, the agent presumably must develop certain subroutines like "do unto others" that are generically useful across different types of games. For instance, in a coordination game the "do unto others" routine might make it possible to alternate between two coordination points, whereas in a Prisoner's Dilemma it could induce limited punishment via Tit-For-Tat-like behavior.

Bednar and Page (2006) have constructed a *games*-theoretic model of culture based on this idea.[5] The model considers six distinct games (including the Prisoner's Dilemma and some variations of the Battle of the Sexes), each of which has a selfish and a cooperative action. Agents play ensembles of these games formed from subsets of the six possible games. The strategies of the agents are based on the evolution of automata framework introduced previously, slightly modified to allow each of the possible games in the ensemble to evolve its own starting state rather than always starting the machine in state 1.

We can think of the number of states in the Moore machine as defining the cognitive capacity of the agents. The number of available states represents a cognitive budget constraint. Just as people have finite wallets, they also have finite brains. To the extent that an agent can share cognitive capacity across multiple games by employing common subroutines, they can do more with less. This implies a cognitive benefit to shared behavior.

When individual games are part of larger ensembles, the presence of the ensemble impacts the strategy for any particular game. For example, if the other games compel the evolution of an elaborate punishment

---

[5] Leady (2006) has considered the problem of when to share subroutines versus when to decompose. Samuelson (2001) has taken on the problem of how to allocate capacity assuming decomposition.

subroutine, then such a routine can be costlessly incorporated into the strategy for yet another game. On the other hand, developing a costly routine for a single game may not be feasible if it does not enhance the agent's strategic ability in some of the other games.

The general result of this research is that context matters: how agents play a particular game depends on the collection of the other games in the ensemble. The ultimate implication of this result is that in worlds in which agents have limited cognitive capacity and face multiple games, we should predict very different behavior than that suggested by the standard game-theoretic models. To put this another way, game theorists have ignored the distinction between partial and general equilibrium analysis that is considered so fundamental in the practice of economics. Agent-based models allow for a natural extension to multiple contexts that takes into account the full cognitive demands placed on agents.

10.5 Evolving Communication

Communication between agents is observed throughout complex adaptive social systems. Communication phenomena occur across a variety of agent types and scales, including biological agents ranging from quorum-sensing bacterial cells to "singing" humpback whales, artificial agents like computer networks, and social institutions like corporations and legal systems. In all of these systems, communication mediates the behavior of the interacting agents, allowing a degree of coordination that can ultimately improve each agent's performance.

Here we focus on the origins of strategic communication. By investigating the origins of communication, we can begin to understand how effective communication can arise endogenously. That is, we would like to investigate worlds in which the meaning of a particular communication is imposed by the decentralized reactions of the other agents in the system, rather than being directed by some central enforcement authority (like a court system or the Académie française). We also want to treat communication as a strategic phenomena, where agents must consider both what they will say to others and how they will react to what others say.

The analysis of endogenous, strategic communication has a variety of uses. First, it allows us to investigate the emergence of communication in adaptive social systems. Second, it gives us a good benchmark for the potential of communication to alter and, perhaps, improve a system's behavior. Finally, it may provide a good technique for imbuing artificial systems, such as computer networks, with the benefits of adaptive communications.

The evolving automata framework discussed previously provides a simple way to model endogenous, strategic communication (Miller et al., 2002). Suppose that there is a known set of communication tokens, $\{C_1, C_2, \ldots, C_N\}$, that can be sent or received by agents. All agents recognize the separate tokens, but no other formal meaning is placed on the tokens. If we allow these tokens to be possible inputs and outputs to an automaton, we can directly apply the evolving automata framework developed earlier.

Miller et al. (2002) apply this idea to the problem of cheap talk in games. Suppose that two agents are about to play a single-shot Prisoner's Dilemma. Prior to playing the game, the agents are allowed to have a "conversation" by exchanging communication tokens. During the conversation, the agents simultaneously send each other symbols until either they both indicate that they have decided on a move in the final game or time has run out. If they both indicate that they have chosen a move, the move is simultaneously revealed and the agents receive the usual Prisoner's Dilemma payoffs. If time runs out, agents that have not chosen a move are penalized.

These conditions form a worse-case scenario for communication. In a one-shot Prisoner's Dilemma, defection is a dominant strategy. Under some conditions, communication could alter the situation sufficiently to allow both agents to cooperate. For example, if communication implied binding promises or was difficult to duplicate, then cooperation could emerge. Unfortunately, in the situation outlined here, none of these conditions hold, and communication is only "cheap talk" that should be ignored.[6] Thus, our a priori prediction would be that no communication will emerge in the model and mutual defection will prevail.

To implement the model, we assume a population of agents that are matched in a round-robin tournament with one another. Each agent's strategy is controlled by a Moore machine. The set of actions available to each agent is given by $\{C \oplus C_0, D \oplus C_0, C_1, \ldots, C_N\}$, where C_i is sending communication token i, and $C \oplus C_0$ ($D \oplus C_0$) indicates that the agent has chosen to cooperate (defect) in the game and sends communication token C_0. When an agent sends C_0, all the opponent can infer is that a choice of either C or D was made. Once an agent enters a state with action $C \oplus C_0$ or $D \oplus C_0$, it remains in that state (and continually sends C_0) until either the opponent also indicates that it has chosen a final move or time runs out. Each state of the automata must be able to respond to inputs from $\{C_0, C_1, \ldots, C_N\}$. As in previous models, we allow the

---

[6]Because this is an evolutionary system, it is likely that some agents will face each other in the future. Nonetheless, given the model's structure, agents would have a difficult time maintaining sufficient information to make the system one of repeated play.

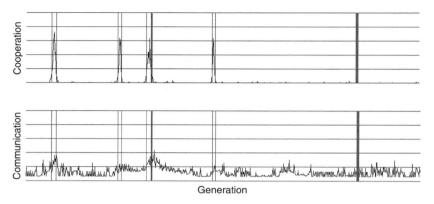

Figure 10.5. Some results from evolving communication. The top panel shows the percentage of total games in which both players cooperated during the experiment (generations are on the x-axis). As seen in the graph, the system experiences occasional outbreaks of mutual cooperation. The bottom panel tracks the average amount of communication tokens exchanged by the players during each generation. Note that communication tends to increase just prior to the cooperative outbreaks and persists for many generations thereafter.

agents to accumulate payoffs from all of their games and then modify the strategies using a genetic algorithm.

10.5.1 Results

The model generates some surprising results (see figure 10.5). Rather than seeing a world filled with mutual defection, we observe occasional outbreaks of cooperation. Observations of the average chat length, that is, the number of communication tokens exchanged before both agents choose final moves, indicate that just prior to the start of a cooperative epoch there is a burst of communication that persists long after the epoch has ended (and only slowly decays back to the low base level).

We can conduct a variety of experiments designed to illuminate the driving mechanism behind the system. These experiments indicate that the system most often finds itself in a situation in which each agent's initial action is $D \oplus C_0$. That is, agents immediately decide to defect and send the signal that they have chosen a final move. If all agents pursue this strategy, we observe a world characterized by no communication and mutual defection.

Every so often, a mutation occurs in which an agent will begin by sending a communication token from $\{C_1, \ldots, C_N\}$ and, if its opponent also sends a communication token from this set, the agent will cooperate;

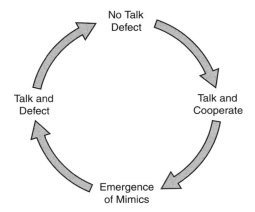

Figure 10.6. Cyclic cooperation under communication. The evolving communication system tends to cycle through four stages. Initially, the world is filled with defectors who do not communicate. Some agents then learn how to communicate and reciprocate communication with cooperation. This leads to the emergence of mimics who destroy the cooperation, leaving the system with a lot of communication and little cooperation. From this state, communication declines and the system relaxes back into the initial state of defection with little communication.

otherwise, it defects. Such agents can survive in a world surrounded by agents doing $D \oplus C_0$, because against these opponents the game ends with mutual defection. If they happen to meet an agent like themselves, however, they cooperate and receive a greater payoff. If two of these communicating agents arise, the conditions are ripe for a rapid spread of cooperation, and we observe a system with lots of communication and mutual cooperation.

Unfortunately, the success of the "communicate and cooperate" agents leads to their own destruction. Evolution creates mimics who send the identical communication stream as cooperative agents, but who then defect. This leads the system into a state where there is a lot of communication, but little cooperation. Eventually, the system relaxes back to the conditions that we described initially, where agents immediately defect without talking. Figure 10.6 illustrates the cycle underlying the system.

Normally, mimicry might be prevented by having an elaborate communication handshake that would be hard to duplicate. Unfortunately, evolutionary systems naturally surmount this obstacle. Suppose we have some agents that develop a very complex handshake before cooperating. The success of these agents will imply that they are often selected for

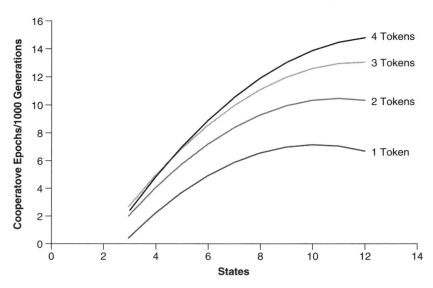

Figure 10.7. Predicted cooperative epochs per 1,000 generations. States gives the maximum number of states each automaton is allowed, and tokens is equal to the number of unique communication tokens (not counting C_0) available to the agents.

reproduction. When they are reproduced, their children will be given the full handshake, and it will only take a single mutation (a change of the final move from cooperation to defection) to create a perfect mimic. This mimic will thrive, because it receives a very high payoff as it defects against its cooperative siblings.

Two parameters of interest in models of strategic communication are the number of tokens available for communication and the amount of processing power in the automata. Based on the experiments described here, we can estimate the impact of these two parameters on the average number of cooperative epochs that arise in the model. Figure 10.7 gives these estimates. Note that as the automata become more powerful (the number of states rises), we tend to see more cooperation. Also, cooperation increases as more communication tokens become available to the agents.

Increases in the size of the automata and number of available communication tokens allow more cooperation to emerge in the system for a variety of reasons. A microlevel analysis of the system suggests that complicated ecologies of strategies develop in the system (see figure 10.8). There is a constant dance among defectors, cooperators, and mimics, and this interplay becomes more elaborate as the processing ability of the

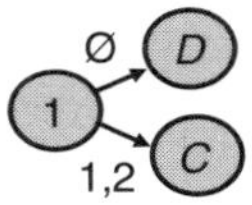

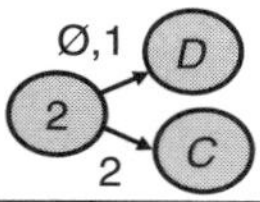

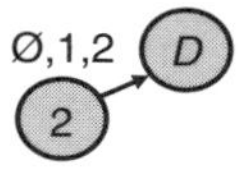

Gen.	% Coop.				
840	0.011	39	4		
841	0.054	31	8		
842	0.097	30	11		
843	0.075	31	10		
844	0.244	15	18	6	
845	0.332	4	17	9	
846	0.327	0	9	17	
847	0.306	1	2	21	
848	0.661	0	4	25	
849	0.673	1	0	24	4
850	0.250	0	0	11	13
851	0.113	1	0	5	23

Figure 10.8. A strategic ecology. In Generation 840 an automaton emerges that communicates by sending C_1 and cooperates with any other agents that communicate (the numbers below each automaton give the size of the respective population). This allows it to cooperate with itself while avoiding falling prey to the predominant population of noncommunicating defectors. In Generation 844 a new strategy arises that communicates by sending C_2 and only cooperates with other agents sending C_2, thus it cooperates with itself while defecting on the previous cooperative strategy (which fails to discriminate among initial signals). By Generation 849 a mimic arises and destroys the value of communication as a cooperative signal. The evolution of strategies in the model allows a complex ecosystem of behavior to emerge. More details can be found in Miller et al. (2002).

automata and the number of communication tokens increase. Under such conditions, agents are able to develop more elaborate handshakes that, while still vulnerable to mimicry, tend to support alternative handshake pathways that can take over once a particular path has been compromised. More elaborate strategies are also able to find and exploit vulnerabilities in the mimics, which also results in longer cooperative epochs.

10.5.2 Furthering Communication

This discussion outlines the impact of communication in the Prisoner's Dilemma. Recent work (Miller and Moser, 2004) used the identical framework but focused on a game of coordination (the Stag Hunt game). Again, the presence of communication fundamentally alters the usual theoretical predictions. In this case, agents are able to coordinate on the superior (yet, previously assumed, impossible to attain) equilibrium point. As before, better communication in terms of processing ability or number of tokens improves the outcome.

In general, communication is capable of productively altering the interactions in a social system for a few key reasons. First, communication expands the behavioral repertoire of the agents, allowing new and potentially productive forms of interaction to prevail. With communication, agents can create new actions that allow them to escape the previous behavioral bounds. The greater the potential of communication, proxied in our discussion by processing ability and tokens, the more possibilities that emerge. Second, communication emerges as a mechanism that allows an agent to differentiate "self" from "other." In the worlds we have explored, agents would like to cooperate in the case of the Prisoner's Dilemma and hunt stag in the case of the Stag Hunt, but the presence (and inherent incentives) of defectors is an ever present danger to adopting such behavior. Communication emerges as a way either to signal a willingness to be nice or to detect meanness. In these systems, this occurs when a fortuitous mutation gives an agent the ability to "speak" and to respond positively to such communication while avoiding harm from those agents that say nothing. By detecting "self" in such a way, the agent can improve its performance even in nasty worlds.

While we see the emergence of communication in this model, note that the agents are preequipped with the ability to send, receive, and process tokens. Thus we might want to investigate the question of how even this more primitive ability to communicate could arise. One hypothesis is that actions, such as the raising of an open hand to signal that "I hold no weapons and do not wish to fight," become "tokenized" over time into pure communication signals, in this case, a gesture indicating hello or "I give up." It might be possible to create a model in which actions with immediate consequences have the potential to devolve into more abstract signals. Another interesting expansion would be to see if grammars can evolve. Grammars use processing to give more meaning to the set of available tokens, and thus we might expect to see the development of grammars in cases where the raw communication needs outstrip the available tokens.

10.6 The Full Monty

We now have tools that permit us to explore systems of interacting, adaptive agents. Within genetic algorithms, we find that details of parameter values or design choices often do not seem to matter in terms of the behavior of the algorithm. Moreover, other adaptive algorithms, like replicator dynamics and neural networks, seem to yield similar results on similar problems. All of this hints at the potential of large equivalence classes interconnecting the space of adaptive-agent models and, ultimately, the potential for a comprehensive and cohesive understanding of how adaptation may drive complex systems.

Agent-based models using artificial adaptive agents offer endless possibilities. At the extreme, we might be able to produce a model of adapting, communicating, multiple-game playing artificial agents that are situated in a world that approximates all of the important social, geographic, and economic systems that we wish to know about. While such a model would be interesting to derive and analyze, by definition it would be big and messy. Indeed, substituting a real world that is tough to understand for an equally confounding artificial one may not be all that helpful. You could argue, though, that the aforementioned artificial world is more easily observed and manipulated than the real one and, as such, might provide a better substrate from which to investigate the world.

The driving ambitions of models that explore game ensembles and communication is the potential to yield insights into phenomena like the evolution of trust and cooperation and perhaps even the emergence of more complex economic and political institutions. Yet, at present we need to keep such ambitions in check so that our reach does not exceed our grasp. These tools make it all too easy to create complicated worlds, but such worlds are typically hard to understand. Fortunately, even in their simplest forms, models of evolving automata offer many opportunities to learn about our world.

CHAPTER 11

Some Fundamentals of Organizational Decision Making

> A life spent making mistakes is not only more honorable, but
> more useful than a life spent doing nothing.
> —*George Bernard Shaw*

ORGANIZATIONS COMPOSED OF COLLECTIONS of agents influence the behavior of systems ranging from biochemical and neurological pathways to political parties and firms. Like all decision-making entities, these organizations must formulate productive actions based on information from their environment. Here, we investigate some fundamental principles underlying this process. We develop these principles by analyzing a model of decentralized decision making that allows us to explore a well-defined class of organizational structures immersed in an ensemble of potential problems. Our goal is to understand better some of the key "natural" constraints governing all organizational systems. The questions we confront include: What are the limits on organizational decision making? Is there a best organizational structure? How do problems influence the types of organizations we observe?

Presumably, we see organizations in the world because of a need to transcend the limits of individual agents. For an organization to be viable, each member must be receiving at least as much benefit from membership as she would from acting alone. Thus, in terms of *benefits*, organizations must be synergistic in the sense that the whole does at least as well as the sum of its parts. This does not mean, however, that organizations must be synergistic in terms of productivity; the minimal requirement here is that the sum must be at least as great as *a* part. For example, in tug-of-war all of the benefit is in winning, and thus adding a person to a team can be beneficial even if the effective pull of the team doesn't increase by the full amount of that individual pulling on her own.

Organizations are able to circumvent a variety of agent limitations. Some organizations are useful because they can aggregate existing characteristics of agents, such as when tug-of-war teams combine each member's strength or schools of fish confuse predators by forming a much larger and more dynamically shaped "individual." At other times, the value of an organization comes through internalizing external benefits, such as flocks of geese (or schools of fish for that matter) having an easier time moving by using the vortices created by other members of the group. Organizations can also allow agents to exploit specialization and circumvent other innate limitations, such as the ability to acquire or access incoming information, or individual bounds on processing the information once it is acquired.

Our focus in this chapter is on how well a given organization can use information from the environment to make "good" decisions. We assume that organizations form to enhance the limited information-processing ability of each individual agent. Our agents will be capable of processing a restricted set of information and then passing on their results to other agents. Eventually, we will require the processing to converge on a single, binary decision. The model is intentionally stark—relying on a very simple notion of information, agent behavior, and organization. Nonetheless, it illustrates a number of more general points.

11.1 ORGANIZATIONS AND BOOLEAN FUNCTIONS

We model organizational decision making by assuming that a group of agents must transform a set of binary information into a binary decision. For concreteness, consider the incoming information as being arrayed on a binary string of fixed length. Organizations must transform this information into a single, deterministic binary choice. For example, if the input consists of two bits of information, the first bit could represent, say, whether the price of a stock was falling (0) or rising (1) and the second bit might indicate whether trade volume was low (0) or high (1). There are four possible two-bit input strings (00, 01, 10, and 11), and a decision rule must associate each of these strings with either a 0 or 1 (which may indicate, for example, whether to sell or buy the stock).

This framework allows us to capture the decision process in a binary rule table that links any potential input string to the appropriate decision. One such rule table is shown in table 11.1, where unless the input string has only ones (say, a rising price and high volume), then the decision is 0 rather than 1 (say, sell rather than buy). Such association tables define Boolean functions.

Thus, each Boolean function defines a "problem" (that is, a mapping from possible inputs from the world to a binary choice) that could be solved by an organization. Therefore the entire problem space confronting these organizations is given by the set of all unique Boolean functions. Here we consider problems that have four-bit inputs. In this

TABLE 11.1
A Sample Rule Table (Boolean Function)

Input String	Choice
00	0
01	0
10	0
11	1

case, there are 16 (2^4) possible input strings, and thus there are 65,536 (2^{16}) unique Boolean functions.

By having such a well-defined ensemble of problems, we can ask questions about the generic properties of organizations as problem solvers. An individual organization's success on a particular problem is given by the number of all possible input and choice combinations that it matches in the associated rule table.[1] In the four-input case, an organization that perfectly solves a problem must match all sixteen entries in the rule table.

To summarize, in the world just described, an organization is any entity that induces a Boolean function (that is, provides a deterministic binary choice when it is presented with a feasible input string). In this chapter we will implement a specific class of such organizations, but many of the arguments we put forth are not tied to this particular form.

Here we consider organizations composed of a set of connected nodes. Each node receives some limited input, either from a string coming in from the environment or from another node. Each node processes this input using a deterministic formula (that may differ across nodes) and emits a single output bit that either flows up to other nodes (as an input for additional processing) or, if it is the topmost node, represents the final choice of the organization.[2]

For example, consider an organization that is composed of three nodes, each of which takes in two binary inputs and emits a single binary output—we label such organizations as $3n2bH$ (3-node, 2-bit, hierarchy). We structure the organization such that the two lower nodes each take in two bits of input from the incoming information string

---

[1] Here we assume that the input combinations arise with equal probability and value. We could adjust performance measures to reflect any biases in either occurrence or value.

[2] The problem outlined here is related to that of integrated circuit design, in which a few simple components must be combined so as to implement key computational functions. The focus of most such work is on implementing designs that give exact solutions to a limited set of functions—here, we are willing to consider inexact solutions to a larger set of problems.

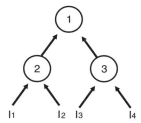

Figure 11.1. A simple organization.

(see figure 11.1). After these two nodes act on their inputs, the single upper node takes the two single-bit outputs from the lower nodes, processes it, and emits the organization's final decision. These $3n2bH$ represent one of the simplest designs capturing the notion of a hierarchically arranged, decentralized organization.

Because each node must deterministically emit a single bit based on two bits of input, the node must embody a two-input Boolean function. Given the four possible combinations of inputs, there are 16 (2^4) possible Boolean functions that could be used for any given node. With three nodes being needed to form an organization, there are 4,096 (16^3) possible organizations of this type (assuming that the lower input connections are fixed).[3]

11.2 SOME RESULTS

As discussed, we view organizations as decentralized collections of processing nodes designed to solve Boolean functions. Like all models, the hope is that this simplification of organizational life will lead to some fundamental insights that transcend the constraints of the model. Our analysis uses computational experiments to provide insight and clarification, linked with more formal mathematical derivations where possible.

The first observation, though simple, has important implications. In our model, organizations always produce an output for any feasible input string. This output is deterministic since, whenever an organization is presented with the identical input string, it always produces the identical choice. Thus, each organization, regardless of its internal

[3] As will become apparent in the results, there are numerous symmetries embodied in these organizations.

design, will always implement a single, fixed rule table. This results in the following:

Claim 11.2.1 *Any deterministic organization will perfectly solve a single problem; that is, it will map every feasible input string to the appropriate output defined by some fixed Boolean function.*

Claim 11.2.1 implies that any given organization will be perfect at solving one particular problem. Later, we use this observation to develop some insights about the relative abilities of large classes of organizations. Note that the claim does not say that any given problem can be solved by a given organizational form, only that a given organization can solve only one problem. It may be the case that more than one organization solves the same problem.

Given a particular organizational form, such as the class of $3n2bH$ organizations, we are interested in the ability of the members of that class to span the potential set of problems. For example, how many of the $65,536$ (2^{16}) possible four-bit input Boolean functions (problems) can be solved by at least one of the possible $3n2bH$ organizations.

We might find that certain organizational forms are inherently limited by their structure in terms of what problems they can solve. Before we investigate the behavior of the $3n2bH$ organizations, we first wish to explore whether there is something inherently limiting about using sets of two-bit input nodes connected in a hierarchy. Here, we note that such hierarchies, as long as we are not restricted to only three nodes and can have multiple connections to the input string, can solve any possible Boolean function:

Claim 11.2.2 *A hierarchical collection of two-input nodes of sufficient size and redundant input connections is able to solve perfectly any possible Boolean function.*

To understand this claim, consider the problem of recognizing a single input pattern. To do this, we could have each of the lowest nodes hooked to the inputs send a 1 if its associated two bits of input match the pattern and a 0 otherwise. If all the nodes above the first level perform an *AND* operation, then the top node in the organization would emit a 1 if the particular pattern was found and a 0 if it was absent. Once we can identify a particular pattern, we can implement any possible rule table by combining these pattern-recognizing hierarchies into a bigger organization. To do so, first create a separate hierarchy for each pattern that results in, say, a 1 in the rule table. Next, combine the output of each of these pattern-recognizing hierarchies with a hierarchy of *OR* nodes, so that the top node of this super organization now emits the appropriate answer to the Boolean function. Thus, given sufficient nodes and careful construction, we can always create a hierarchical organization composed of these types of limited ability agents that can solve any potential problem.

The proof of claim 11.2.2 suggests that we can put an upper bound on the minimal number of nodes needed to implement any Boolean function of a given size by calculating the minimum number of nodes needed to implement this scheme. In the case of four-bit inputs, a hierarchy of three nodes is sufficient to recognize any *specific* four-bit pattern. For a given rule table, we need to identify only the patterns associated with either all of the 0s or all of the 1s in the table. To conserve on nodes, pick the smaller of these two sets, which must be of a size less than or equal to eight (half of the sixteen possible patterns in the table), so we will need at most twenty-four (8 × 3) nodes to recognize the key patterns. The eight outputs from this part of the organization can be combined using seven additional "OR" nodes to give the final answer, implying that we need at most thirty-one nodes to implement any possible four-bit rule table.

We now turn to analyzing the properties of our $3n2bH$ organizational form. Intuitively, it might seem that the straightforward structure of these organizations and the simple problem domain (four-bit inputs) that they confront should make such organizations strong candidates for solving any potential problem in the domain; however, this is not the case. Recall that there are 65,536 (2^{16}) possible four-bit input Boolean functions (problems). Because there are only 4,096 (16^3) possible $3n2bH$ organizations (and from claim 11.2.1 we know that each of these can solve only one problem), then *at most* only 6.25 percent of all possible four-bit problems can be solved by these types of organizations.[4]

The 6.25-percent figure assumes that each organization solves a unique problem, but since more than one organization can solve the same problem, it represents an upper bound on the performance of

[4]One caveat on the previous result is that we do not allow the lowest-level input connections to relocate on the input string. Such simple reorganization might allow an organization to solve previously unsolvable rule tables. For example, a rule table that maps 1000 and 0010 into 1 and everything else into 0 cannot be solved by our $3n2bH$ organizations. However, rearranging the second and third connections from the lower-level nodes results in a rule table that maps 1000 and 0100 into 1, a problem that can be solved. The general issues that arise in this discussion are interesting for organizational theory. For example, Simon's (1969) ideas of decomposability could be applied as some rearrangements of the input nodes transform the problem into something that is easily decomposable. One could also apply ideas on how combinations of "views" of the problem (each node's input connections) and problem-solving heuristics (the actual function each node uses) alter an organization's problem-solving ability.

TABLE 11.2
Problem Solving for $3n2bH$ Organizations

Number of Problems	Number of Organizations Solving Each	Total Organizations
65,016	0	0
490	4	1,960
28	44	1,232
2	452	904

$3n2bH$ organizations. Computationally, we can take each of the 65,536 possible problems and determine the number of organizations that can solve each one. Table 11.2 summarizes these results. Note that 65,016 problems cannot be solved by any of the possible organizations—that is, only 0.79 percent of all possible problems can be solved by this class of organizations. Of the 520 problems that can be solved, the vast majority of them (490 or 94 percent) are each solved by four organizations.[5] Two problems, those with rule tables of either all zeros or all ones, are solved by 452 organizations each. Much of the redundancy of these latter two problems arises because there exists a variety of ways to configure some key nodes and force the appropriate solution on the organization.

As the input stream gets larger, we find that the potential problem space covered by two-bit, nonoverlapping, hierarchical organizations rapidly goes to 0. An organization with x levels in the hierarchy requires $x(x + 1)/2$ nodes and can take in inputs with $2x$ bits. We therefore have $16^{x(x+1)/2}$ possible organizations that must embrace $2^{2^{2x}}$ problems. Even for small values of x, the ratio of possible organizations to problems quickly goes to zero.

11.3 Do Organizations Just Find Solvable Problems?

The preceding analysis suggests that classes of organizations may be severely constrained in their ability to solve all possible problems. While at some level this observation seems sensible, it does potentially challenge the more commonly held belief that organizations arise to solve whatever problems they may confront. Perhaps organizations are much more

---

[5] Since two bits are passed up from the lower nodes to be processed by the upper one, we can always put in an alternative function in either (or both) of the lower nodes that results in the opposite output (that is, sends up a 1 instead of a 0 and vice versa) and then compensate for this change by altering the upper node's function. Thus, for any solvable problem, there are at least four possible organizations that can solve it.

constrained than we have previously imagined, and rather than having the ability to solve any possible problem that comes their way, they instead can only exist in worlds that embody the right kind of solvable problems.

Even a generic organization, with some simple modifications in the spirit of the constructions outlined in claim 11.2.2, might be able to solve a larger set of problems. However, even if organizations are adept enough to implement such changes, it is likely that such modifications are rather costly and brittle once an organization adds more than a few defaults to its generic design.

Alternatively, it might be the case that even though the problem space is vast, by chance the actual set of productive problems confronting organizations happens to fall within the small subset of problems that are solvable by generic organizations. Of course, the probability that time after time, the problems that need to be solved by different organizations happen to be "easy" would appear to be vanishingly small, and it would be hard to support this chance-based hypothesis.

A better resolution of these issues is to admit the possibility that the organizations that we observe in the world are more driven by the set of solvable problems than is usually assumed. That is, organizations can arise only in those very narrow niches that contain easily solved problems. This Goldilocks view of the world implies that organizations emerge only when the conditions are just right, and suggests that focusing on the narrow spectrum of solvable problems may provide a good purchase upon which to build our understanding of organizations.

In table 11.3 we provide the distribution of four-bit problems that can be solved by at least one member of the class of $3n2bH$ organizations. Note that $3n2bH$ organizations are most successful on the extremes of the functions, that is, on those parts of the problem space where there are very few (or very many) inputs that require the same choice. Such a relationship is suggested by the proof of claim 11.2.2, as the more lopsided the function, the fewer the patterns that need to be recognized, leading to an easier calculation. Since the majority of the Boolean functions exists in the range with an intermediate mix of output values, most will not be solvable by our organizations.

11.3.1 Imperfection

The previous discussion focused on the conditions necessary for an organization to perfectly embody a given rule table. Of course, in many systems having answers that are close is often good enough, so investigating less than perfect performance is of interest. As we saw,

TABLE 11.3
Problem Distribution for $3n2bH$ Organizations

Number of 1s in the Function	Number of Such Boolean Functions	Number Solvable	Percent Solvable
0	1	1	100.00
1	16	16	100.00
2	120	48	40.00
3	560	32	5.71
4	1820	44	2.42
5	4368	0	0.00
6	8008	64	0.80
7	11440	16	0.14
8	12870	78	0.61
9	11440	16	0.14
10	8008	64	0.80
11	4368	0	0.00
12	1820	44	2.42
13	560	32	5.71
14	120	48	40.00
15	16	16	100.00
16	1	1	100.00

one case where less than optimal performance is interesting is when a problem cannot be perfectly solved by any organization in a given class. We could also consider the case where an organization must confront an ensemble of problems, and since an organization can only perfectly implement a single such problem, we need an expanded notion of performance across the ensemble. We can measure the performance of any given organization by the number of inputs it gets correct in the function—what we will call accuracy. Intuitively, it might seem that certain organizational forms are superior to others a priori. However, we note the following:

Claim 11.3.1 *Without reference to a particular problem, all deterministic organizations are equally accurate on Boolean problems.*

The proof of this proposition is straightforward. Consider the class of n-bit Boolean functions. There are 2^n entries in the input domain. Since any organization operating on this domain is able to implement perfectly a single, Boolean function (claim 11.2.1), there are exactly $\binom{2^n}{b}$ other rule tables that differ by b bits from the one generated by the organization. Thus, this organization will be b bits off on $\binom{2^n}{b}$ other problems. This argument does not depend on the specific Boolean function implemented

TABLE 11.4
Problem Accuracy on Four-Bit Inputs, Where Accuracy Is the Number of
Rule-Table Inputs (Out of Sixteen Possible) That It Gets Correct.

Accuracy	Number of Problems
0	1
1	16
2	120
3	560
4	1820
5	4368
6	8008
7	11440
8	12870
9	11440
10	8008
11	4368
12	1820
13	560
14	120
15	16
16	1

by the organization, so any organization will have an identical accuracy distribution. Of course, the claim does not state that organizations must have the same accuracy on the same problems—each organization may miss different problems—but only that the distribution is identical. For example, in a world with four-bit inputs, *any* deterministic organization embodies the accuracies shown in table 11.4.

Claim 11.3.1 suggests that there is no best organizational design without knowing about the problems that will be confronted. In essence, any organization is equally good at solving problems, broadly defined.[6] While the notion that particular problems require particular solutions is not too surprising, the observation that there is a much deeper level of equality across all organizations is less obvious.

While all organizations are equally imperfect across all problems, good organizations are those that produce the proper answers to the problems that the organization actually confronts. If we know in advance which problems need to be solved in the world, we could rank organizational forms by their ability to provide accurate answers to the particular ensemble of problems in play. There may be an interesting interplay

---

[6]This is a similar conclusion to the No Free Lunch result of Wolpert and Macready (1997).

between "generalist" organizations that solve a set of problems relatively well and "specialist" ones that have higher accuracy on a narrower set of problems in such a world.

11.4 FUTURE DIRECTIONS

The view of the world embodied in the preceding model and analysis is a simple one with many avenues remaining for investigation. It begins with the premise that organizations must make decisions based on information they receive from the environment. By introducing the notion that problems are in the form of Boolean functions, we can exploit the resulting structure both computationally and mathematically.

Any deterministic organization, regardless of its internal structure, will be able to perfectly implement a single Boolean function, that is, be able to solve a single problem. Of course, nothing prevents more than one organization from solving a given problem, and large overlaps may be the norm. The idea of "one organization, one function" has some powerful implications; for example, it allows us to quickly determine the potential coverage of a given class of organizations across the problem space. Indeed, for certain classes of organizations, it is likely to be the case that the problem space quickly overwhelms the solution ability of the underlying organizations. That is, organizations may be productive on only a small set of all possible problems. This may imply that organizations may be more reactive than proactive, working well only when the problems are easily solvable rather than solving whatever problems come their way.

The fact that each organization implements a single Boolean function also implies that there is a deep parity among all organizations in terms of problem-solving ability. This parity suggests that all organizations exhibit identical generic performance characteristics. Thus, in this sense the most elaborate organizational form imaginable is no better than an organization that always makes the same choice, regardless of input.

The simple model developed here begins to highlight some of the fundamental constraints on organizational behavior and should serve as a productive basis for further investigations. The results, while perhaps obvious a posteriori, challenge existing wisdom and give a new emphasis on aspects of the problem that were previously relegated to minor roles. It is likely that further analysis of these simple systems will yield important insights about organization in both natural and artificial systems.

PART V

Conclusions

CHAPTER 12

Social Science in Between

> The machine does not isolate man from the great problems of nature but plunges him more deeply into them.
> —*Antoine de Saint-Exupery, Wind, Sand, and Stars*

> If you want to build a ship don't herd people together to collect wood and don't assign them tasks and work, but rather teach them to long for the endless immensity of the sea.
> —*Antoine de Saint-Exupery, Wisdom of the Sands*

> Our inventions are wont to be pretty toys, which distract our attention from serious things. They are but improved means to an unimproved end.
> —*Henry David Thoreau, Walden*

HERE WE DISCUSS THE IMPACT of complex adaptive systems on the social sciences. Our book's central theme, "The Interest in Between," provides a framing for this discussion. The complex adaptive social systems view of the world allows us to explore the spaces between simple and strategic behavior, between pairs and infinities of agents, between equilibrium and chaos, between richness and rigor, and between anarchy and control. These spaces lie between what we currently know and what we need to know. They are not subtle refinements on the landscape of knowledge but represent substantial deviations from what we typically assume. The story is told of a geologist who walks to the rim of the Grand Canyon and remarks "something happened here." Social scientists seem to be haunted by their own canyons, and it is time that we actively engage these mysteries and begin to explore them.

The social sciences have pursued a variety of methodologies. Techniques like empirical research, natural and laboratory experiments, historical investigations, qualitative methods, mathematical and game theory, and computational models have all been used. In some cases, these methods have been deployed and refined by thousands of scientists over many, many decades. In other cases (like computational models), they have been used by just a handful of scientists only recently. Each approach can be both a complement and substitute for the others.

Thus, careful empirical work can both substitute for, and complement, laboratory experiments; computational models can enhance, or replace, mathematical ones; and so on.

In the absence of any one method or idea, science would continue to advance, albeit perhaps at a slower pace or in a different direction. Nonetheless, sometimes the changes in the pace and direction brought by a new methodology or set of ideas can be significant. In this chapter, we outline some initial contributions we attribute to the complex adaptive social systems view of the world. We also highlight some of the new frontiers that can now be explored—the interest in between the usual boundaries.

12.1 Some Contributions

It is still too early in the development of complex adaptive social systems ideas to fully assess their contributions. We know that some of the results that have been found can be replicated using more traditional techniques, though it is often the insights and discoveries made with the new methods that allow the old ones to be applied. Ultimately, the complex adaptive systems approach has focused our attention on new possibilities. Even though the applications of these ideas are still in their infancy, they have already begun to contribute to our understanding of key social processes.

A key contribution of complex systems has been a better appreciation of the power and mechanism of emergence. Models of self-organized criticality show how systems can locally adapt to a critical region in which the global properties of the system take on regular behavior, such as a power-law distribution of event sizes. Such ideas are likely to serve as fodder for explaining various social scaling laws, like the distribution of incomes or firm sizes (Axtell, 2001).

Perhaps many features of social systems are the result of self-organization. Computational models of market behavior have highlighted key features that allow the emergence of predictable prices and trading patterns in markets (Rust, Miller, and Palmer, 1992, 1994; Gode and Sunder, 1993). In particular, this work has shown that a sufficient requirement to see such behavior emerge is the presence of simple institutional rules that force new offers to better existing ones. Such an insight radically altered the existing view—one that relied on the innate cleverness of self-interested traders—of the driving force behind Smith's invisible hand (see figure 12.1). The emergence of organization via decentralized means is apparent in the example of voting with your feet explored in chapter 2.

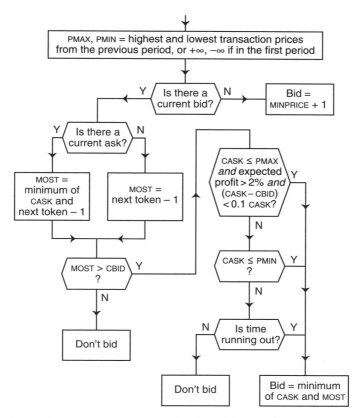

Figure 12.1. Simple trading strategies dominated the Double Auction
Tournament (see Rust, Miller, and Palmer, 1994). Notwithstanding the presence
of some very complicated strategies based on various economic and statistical
theories of trading, it was the simple strategies that won the tournament.
Depicted is a schematic of Kaplan's winning strategy. It allowed other traders to
submit bids and asks, and took advantage of any profitable opportunities when
the spread between the bids and asks was small. This strategy was an
"information parasite" that fed off of the actions of the other agents in the
ecosystem.

Models of emergence also provide insights into the robustness of the
underlying system, as the essence of emergence requires entities to be
able to maintain their core functionality despite what are often radical
changes from both within and without. Using emergence ideas, we can
begin to understand the robustness of systems such as markets, cultures,
and organizations like firms and political parties.

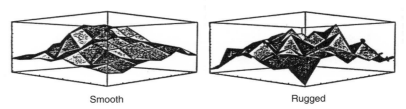

Smooth   Rugged

Figure 12.2. Two landscapes generated by nonlinear interactions. As nonlinear interactions increase in a system, the numbers of peaks and valleys increase as well and the landscape becomes more rugged. Agents with limited search abilities can get trapped easily on local optima when the underlying landscape is rugged. Landscape models have been used in the social sciences to study topics ranging from politics to technological innovation.

Another contribution of complex adaptive social systems has been a recognition of the importance of nonlinearities and interactions. To take one example, consider agents that must blindly search across the world to achieve some goal. To keep such models mathematically tractable, we often need to assume that agents are completely blind (and hence just randomly search), are completely omniscient (making search trivial), or exist in a smooth and single-peaked world (where groping results in optimality). All of these assumptions are both unsatisfying and unrealistic. The complexity approach considers landscapes in which the various elements of the space interact in nonlinear ways, resulting in a convoluted world with many peaks and valleys (see figure 12.2). Once agents are placed in such a world, a whole new realm of behavior opens up. Agents find themselves in a path-dependent world, in which early choices determine future possibilities (Page, 2006). Tipping points and critical junctures emerge, where a given system can rapidly change its characteristic behavior.

The notion of search across a rugged landscape provides a new purchase from which to consider ideas like innovation and political platform formation. For example, we can model firms competing against one another to develop good technologies, where a given technology is described by, say, a binary string in which each bit encapsulates some technological feature that interacts with the other bits (the wing shape of an airplane interacts with its power plant choice, which interacts with its fuselage materials, and so on). Now the process of technological invention becomes a search problem across a rugged landscape, where past triumphs and new discoveries form the basis of new technologies that are brought to the market.

The new network theory has also been a major advance facilitated by the complex systems approach (Newman, 2003). While networks—and, more important, the interactions among agents they facilitate—have long been considered by social scientists, especially sociologists, a wave of recent interest has been prompted by computational and mathematical models created by complex system researchers. Rather than focusing on any particular network, this new work considers the generic properties of social connections. Computational modeling allows researchers to create massive numbers of networks that share particular connectivity patterns, and from these derive generic patterns of behavior. These same researchers have begun to mine new sources of on-line data, providing new examples of networks that heretofore would have been impossible to collect and analyze.

Complex systems ideas have also led to new advances in the modeling of adaptation. Adaptive agents can often radically alter the behavior of our models. For example, consider the formation of political platforms by competing parties. If the parties are able to optimize with perfect knowledge, then we predict that incumbents always lose elections and the party platforms we observe will forever follow a chaotic path. Under adaptive agents (see figure 12.3), the platform dynamics behave in a way that is much more consistent with the real world—they slowly converge to good social outcomes that can be tied to the underlying preferences of the voters (Kollman, Miller, and Page, 1992). Moreover, incumbency advantages spontaneously arise due to the inherent search problems faced by adaptive parties. In such models, the search landscape of each party is coupled to those of the other parties, and the landscapes dance around with one another as one party alters its platform in response to platform changes made by the other parties.

Computational models have opened up vast new frontiers for exploring the learning behavior of agents. To take one example, consider learning in games. The last half of the twentieth century witnessed a tremendous intellectual effort aimed at refining various game solution concepts. Toward the end of this period, good experimental data on how agents actually played games began to emerge, and it was found that many of the formal solution concepts failed to predict what was happening in the experiments. Over the past decade or so, computational learning models have arisen to explain the divergence. For example, Andreoni and Miller (1995) showed how a simple model of learning based on a genetic algorithm can be used to reconcile differences between the theoretical and experimental results arising in various auction markets (see figure 12.4).

Similarly, computational models have played a pivotal role in illuminating issues surrounding the emergence of cooperation. For example,

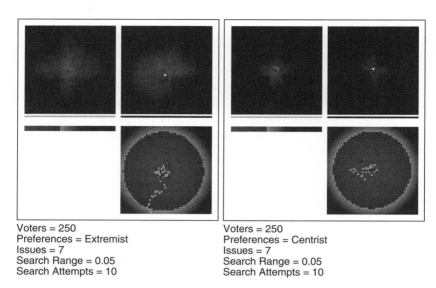

Voters = 250
Preferences = Extremist
Issues = 7
Search Range = 0.05
Search Attempts = 10

Voters = 250
Preferences = Centrist
Issues = 7
Search Range = 0.05
Search Attempts = 10

Figure 12.3. Political landscapes and platform search. The political landscape facing a party is tied to the preferences of the underlying voters and the position of the opponent. When voters have extreme preferences (*left half*), the landscapes facing each party become more rugged and diffuse (*upper panels*), while under centrist voters (*right half*) they become much more concentrated. In both cases, the platforms of adaptive parties tend to converge on good social outcomes (*lower right of each diagram*).

Axelrod's (1984) landmark study relied on a tournament of computerized strategies to investigate strategic behavior in the Prisoner's Dilemma game, and Miller (1988) showed how such cooperation can emerge among adaptive agents. Work is also ongoing that incorporates processes of social learning whereby agents learn by observing others (see, for example, Vriend, 2000).

12.2 THE INTEREST IN BETWEEN

The preceding discussion provides a few examples of where the complex adaptive social systems approach has made contributions to advancing the frontiers of the social sciences. While dwelling on past accomplishments is useful, we are more interested in the future opportunities that are potentially available. The study of complex adaptive social systems opens up vast new frontiers in the social sciences. These frontiers exist in the space between the current boundaries imposed by traditional ideas and methods.

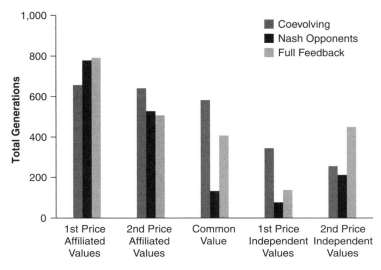

Figure 12.4. Coevolution and learning in auction markets. Here, artificial adaptive agents "learned" to bid in various single-sided auction markets using a genetic algorithm. The patterns exhibited by these artificial agents paralleled those observed in laboratory experiments with humans. Note that agents that coevolved with other learning agents spent more time using optimal bidding strategies (y-axis) than agents that learned in an environment populated by expert strategies. More details can be found in Andreoni and Miller (1995).

12.2.1 In between Simple and Strategic Behavior

Consider a simple game like tic-tac-toe (aka noughts and crosses). Few adults actively play tic-tac-toe, as after a fairly short learning period almost anyone figures out how to force the game to end in a draw and thereafter it is not of much interest. Technically, tic-tac-toe is a sequential game with perfect information. Such games can be solved by mapping out all of the possible paths of play and then working backward through the resulting tree and selecting moves that will force the play of the game down favorable paths. Even though the possible paths in tic-tac-toe are enormous, some symmetries in the game that can be exploited (for example, the nine possible first moves can be collapsed down to three) allow adults to intuit the game tree and do the necessary backward induction without too much effort. Indeed, even chickens can be trained to play the optimal strategy (Stuttaford, 2002).

Although adults do not get much joy out of tic-tac-toe (other than playing games against chickens), children—who are unable to do the necessary calculations—can enjoy the game for hours. Lest adults feel too superior to their children, adding even slightly more complication can quickly overwhelm our own cognitive abilities. Three-dimensional tic-tac-toe might qualify here, as would the game of chess.

Chess has fascinated players in its modern version for over half a millennium. Alas, it is a simple game, with sixty-four squares, sixteen pieces of six possible types on each side, and a limited set of movement and engagement rules. Nevertheless, it has generated a vast literature, rancorous scholarly debates, challenging philosophical quests, and the occasional international incident. The odd thing is that chess is identical to tic-tac-toe, in that it has a well-defined game tree that, in theory, we could work our way through and develop an optimal strategy. If our cognitive abilities were just a bit higher, all of the fuss about chess might be a bit embarrassing (to put this in perspective, imagine "Fischer versus Spassky, the tic-tac-toe match of the century").

The strategic space between tic-tac-toe and chess is an interesting one. On one hand, both games are isomorphic and, in a very real sense, trivial to play. On the other hand, while this statement has some meaning for tic-tac-toe (at least for adults), it seems rather empty for chess. Although we could assume the existence of a chess god, through which chess becomes a trivial game using backward induction, such an approach yields little insight into how chess is really played by humans. While toward the end of a chess game we may indeed fall onto an equilibrium path of play, most of the game is played in a wilderness far from any known equilibrium.

Recent developments in computerized chess programs are instructive in terms of the interest in between simple and strategic play. Like humans, computers are unable to generate the entire game tree for chess except toward the end of the game. Therefore, programs must rely on various heuristics (for example, queens are more valuable than rooks), calculations of localized portions of the game tree (often using clever pruning to avoid pursuing likely dead ends), and other means to decide on their moves.

Social science has struggled to come to grips with how to model human behavior. Simple behavioral rules such as price-taking behavior and voting along party lines dominated social science a half century ago. Then, the tide turned toward models that relied on rational actors who were able to do extraordinary calculations on simple problems. More recently, we have seen a movement toward behavioralism and learning models. At each point along this path, social scientists have struggled with what to assume about behavior. A complex adaptive systems approach allows the level of agent sophistication, and even the behavior itself, to adapt. The appropriate level of strategic behavior is not always clear, as we might expect people to be strategic in some contexts and rule following in others. Nonetheless, we have good evidence that humans do not always act like rational agents and that adaptive behavior may lead to very different outcomes, and thus we need some flexibility to be able to explore the interest in between the strategic extremes we have come to rely on.

12.2.2 In between Pairs and Infinities of Agents

Most social science models require either very few (typically two) or very many (often an infinity) agents to be tractable. When an agent interacts with only a few other agents, we can usually trace all of the potential actions and reactions. When an agent faces an infinity of other agents, we can average out (in physics-speak, take a mean field approximation of) the behavior of the masses and again find ourselves back in a world that can be easily traced. It is in between these two extremes—when an agent interacts with a moderate number of others—that our traditional analytic tools break down.

Unfortunately, most economic, political, and social interactions involve moderate numbers of people. Sometimes two firms do compete for a single account, but more often than not dozens of firms compete for dozens of accounts simultaneously. Once we find ourselves in such a world, our traditional analytic tools fail us. Of course, notwithstanding the futility of our tools, actual firms do continue to operate in such contexts, so there must be some mechanisms, albeit imperfect ones, that come into play and allow firms to survive. Similarly, the world of politics is not fully captured by either two-person or large population games. While we do see two candidates squaring off in an electoral battle, this is typically the exception rather than the rule. A United States senator interacts with ninety-nine other senators. To be effective, senators must navigate a vast strategic landscape that involves voting, amendments, interest groups, lobbyists, constituents, bureaucrats, and other branches of the government. Perhaps some of these domains can be isolated and distilled to interactions with only a few or infinitely many other agents, but such an approach quickly succumbs to the reality of the situation. Moreover, even when the interaction is limited to one dimension, it is difficult for the repercussions to be fully isolated. Almost all actions taken by an agent have implications across many games simultaneously, and even if each of these games has a single opponent, the constellation of them does not.

As we start to increase the number of agents we consider in a model, the mechanisms facilitating the interactions among agents become important. One way to keep things tractable is to assume that agents exist in a soup and randomly pair off with one another for an occasional clash. Models of the spread of disease often make this type of assumption. Alternatively, we can assume that everyone interacts with everyone else simultaneously. General equilibrium market models and political models often make this assumption.

New modeling techniques, combining both mathematics and computation, allow us to make the more realistic assumption that social activity takes place in between these extremes. In these models, agents interact with one another over well-defined networks of connections; for example, diseases are transmitted because two people share the same place of work or travel via the same airline hub and agents trade with one another because they find themselves in the same marketplace (whether this is a city on an ancient trade route or an online auction).

Moving in between the old boundaries alters how we think about, and attempt to change, the world. For example, previous disease models assumed random mixing and were solved using a system of coupled differential equations. Although random mixing may be a good assumption if we are modeling the spread of a cold in an elementary school classroom, it is much less useful if we are trying to model the spread of a sexually transmitted disease such as HIV-AIDS. The assumption of widespread promiscuity that knows no geography (random mixing) fails to appreciate the reality of sexual contact structures. When such contact structures are explicitly incorporated into the model, we get more accurate predictions and better policy prescriptions.

12.2.3 In between Equilibrium and Chaos

The rise of complex adaptive systems and its core ideas stems partly from the intrinsic power of the metaphor. If you consider the data from key political, social, and economic processes, it is not clear whether equilibria are the exceptions or the rule. Stock markets soar and crash (LeBaron, 2001). Political parties rise and topple (Jervis, 1997). Terrorist acts emerge from, and are perpetuated by, loose networks. While the notion of social equilibria is an important one, and perhaps even these phenomena are best reflected as a series of (apparently rapidly changing) equilibria, we may need to go beyond equilibria to truly understand the social world.

Complex adaptive systems models allow us to explore the space between equilibrium and chaos. In the starkness of neoclassical models, exchange markets result in a single, stable price equating the quantity supplied with the quantity demanded. Unfortunately, our experiences with real, experimental, and artificial markets indicate that the actual behavior of a market is not so easily captured. In real markets phenomena like clustered volatility and excess trading remain difficult to explain, in experimental markets traders seem to be less strategic and far more irrational than expected, and in artificial markets even minimally rational traders cause the market to achieve high levels of ex post efficiency, even though the observed price path is very noisy.

The equilibrium predictions of the standard market model in economics contrast sharply with those of spatial voting models from political science. With even minimal complication, spatial voting models rarely have equilibria (Plott, 1967). Yet, political parties do seem to demonstrate a fairly high degree of stability on many issues. As previously mentioned, in a model using adaptive political parties, parties tend to converge and dance around the social center of the policy space (Kollman, Miller, and Page, 1992). This latter result is related to the coupled landscape metaphor we discussed earlier. Consider a landscape where the coordinates are positions on policy issues and the height gives the number of votes such a platform would receive. Adaptive political parties move around such landscapes in search of the (metaphorical) high ground. As one party alters its policy positions, however, the landscapes of the other parties are changed. Thus, the political process is one in which parties must actively seek the high ground, even as the landscape underneath them constantly undulates. Although such a process has the potential to generate a collection of aimlessly wandering parties, we find that most of the time the high ground, while ever changing, tends to be concentrated in a contained region of the policy space resulting in relatively stable platforms.

Equilibria, when they exist, are an important organizing force in social systems. Nonetheless, there is no a priori reason to think that equilibria must exist. If we want to understand social systems, we must also account for those that are complex. As shown by the spatial voting model, the lack of equilibria does not necessarily mean a lack of predictability and insight. Using the techniques of complex adaptive social systems, we now have the capability to explore those systems that lie in between equilibrium and chaos.

12.2.4 In between Richness and Rigor

Early proponents of complex adaptive social systems models were optimistic about the prospects for using these models to combine the richness of more qualitative methods with the rigor of mathematics. Qualitative methods provide great flexibility in terms of the types of problems that can be analyzed. At the same time, these methods are often vague, inconsistent, and incomplete. Mathematical methods tend to be more rigorous with exacting notions of how models are formed and solved. Yet, the cost of this rigor is often a loss of richness in what can be studied. Complex systems models may be able to bridge the gap between richness and rigor.

Consider the problem of getting people seated on a commercial airplane. Airlines can realize considerable savings by reducing boarding times because with faster boarding they can fly the same number of routes with fewer planes. Suppose we have a group of, say, one hundred people waiting in the passenger lounge that we need to get seated as quickly as possible on the waiting aircraft. Passengers must board the aircraft, travel down a lone aisle that is easily obstructed by other passengers, stow any baggage, and get to their seat and sit down. The only real control the airline has over this process is the order (based on seat assignments) in which it allows the passengers to board. A very common system in current use is to allow passengers to enter the plane starting at the rear of the aircraft and moving forward, but a number of alternatives exist, including allowing window-seat passengers to board first, alternating between the two sides of the aircraft, and so on.

We can construct a model of this process in a variety of ways. One approach would be to use the average time it takes a passenger to walk, stow baggage, and get seated, and from this develop a mathematical queuing model. As an alternative, we could incorporate much more fidelity into the model via an agent-based model, in which passengers have connections to one another (say, business travelers versus families), alter their behavior in response to other passengers (stow their bags up front if they cannot immediately get to their seat), and so on. Even if we use an agent-based model, we still must decide on how much detail to build into the model. At one extreme the model would look very much like a mathematical queuing model (with the only difference being that we are using the computer to solve it rather than formal equations), whereas at the other it could be a very detailed simulation of every aspect of the passenger experience.

The agent-based model will be much messier than the one that relies on gross averages. Given that we strive to have stark models, this is a disadvantage. Yet, we also strive to have useful models, and depending on the questions we wish to tackle, we need to be willing to trade off starkness for usefulness. Through stark models we can develop broad intuitions. Through empirical analysis and case studies we can get very detailed accounts of what happens under exacting circumstances. Rich computational models allow us to explore the delicate interactions inherent in a system in a much more expansive way and fill in the space in between.

12.2.5 In between Anarchy and Control

The stock market exemplifies the space between anarchy and control. Our theorems tell us that the market should efficiently aggregate information through the price mechanism. Yet, fluctuations in price appear to far outstrip variations in information. The market sometimes appears to have a mind of its own, yet it does not collapse into complete anarchy. Computational models allow us to mimic such processes (Arthur et al., 1997). They produce behavior not unlike real markets, and we can use them to begin to experiment with attempts to control such worlds. For example, we can see if increasing the amount that can be bought on margin will reduce or eliminate bubbles.

We can extend this idea to think about institutions more broadly. Attempts to assist developing countries through institutional reforms and large projects have, on the whole, been unsuccessful (Lewis and Webb, 1997). People who study development have learned that it may be difficult to find a common method that works across all environments. An institution that works in one culture may not work in another. Ostrom (2005) explains these differences by reference to context. Institutions do not sit in isolation from one another, but are linked to each other and the culture within which they exist. Cultural features like the level of trust, the set of common behavioral rules, and the density of social networks all provide an important context for an institution (Bednar and Page, 2006). We can use computational models to explore these contexts and develop appropriate institutional designs.

Harnessing emergence may be an important means by which to create institutions that can use apparent anarchy to create control. As we saw in chapter 2, a well-designed political institution can introduce noise into a decentralized system in such a way that it promotes the emergence of productive global organization. We also know that institutions like markets can be effectively used, say, to aggregate opinions about political races and world events. We suspect that complex systems ideas will lead to a new appreciation of the importance, and potential for exploitation, of the space between anarchy and control.

12.3 Here Be Dragons

The complex adaptive social systems approach provides many opportunities to explore the interest in between the usual scientific boundaries. This vast unexplored territory is home to many of the most interesting and ultimately important scientific questions. Nevertheless, we have tended not to stray too far from known waters for fear that *hic sunt dracones*.

We now have within our grasp the ability to explore these uncharted waters. Like any such exploration, perils abound. It may be that our intellectual conveyances are inadequate to the task, and that we will founder upon the many shoals that surely exist beneath the inviting seas. Or, perhaps this territory is one of false promises, and our explorations will uncover little of value. Nonetheless, the early expeditions prove that the seas can be sailed and suggest at least the possibility of potential riches, so explore we must, even if, as T. S. Eliot (1942) wrote, "the end of all our exploring will be to arrive where we started and know the place for the first time."

Epilogue

It is all very beautiful and magical here—a quality which cannot be described. You have to live it and breath it, let the sun bake it into you. The skies and land are so enormous, and the detail so precise and exquisite, that wherever you are, you are isolated in a glowing world between the macro- and the micro- where everything is sideways under you, and over you, and the clocks stopped long ago.
—Ansel Adams, Letter to Alfred Stieglitz

Do not seek to follow in the footsteps of the wise. Seek what they sought.
—Matsoo Basho

HERE WE HAVE EXPLORED MODELS that from simple beginnings result in systems imbued with scientific beauty and mystery. The precise and simple details of each model are fully responsible for, yet simultaneously rather removed from, the majesty of the resulting outcomes. We find ourselves immersed in a world that lies between the micro and macro, that twists our experiences and expectations, and that hints at the interest in between the usual boundaries we impose on our models. We are on the edge of a vast frontier, where the exquisite composition of what once was a distant world is beginning to slowly yield its secrets to new ways of inquiry that are starting to make the faraway nearby.

THE INTEREST IN BETWEEN

The study of complex systems attempts to illuminate the interest in between our usual scientific boundaries, and in so doing, paradoxes abound. It is the study of how the few are different from the one or the infinite. It is the exploration of time in a highly choreographed dance. It is a search for tight connections in a loosely coupled world. It is the precise characterization of when details do not matter.

Modeling, by its very nature, is about extremes. When we model systems we attempt to push our scientific fantasies to the farthest edge of reality, in hopes of gaining some new insight. Unfortunately, sometimes in the pursuit of extremes we kill off the most interesting parts of the world—"water which is too pure has no fish."

One important insight from models of complex adaptive social systems is the interest in between the extremes. Using these models, we are finding that as we move away from the extremes we do not incrementally approximate what has come before, but instead are thrust into new realms of experience.

The interest in between has long been known. For example, in economics we have lovely, compact models of firms behaving as monopolies, duoplies, and perfect competitors, but once we are in the realm of a *few* firms, modeling and prediction become difficult; in physics we can solve mathematically the two- and ∞-body problem, but no clean solutions exist for the intermediate cases. It is as if much of modern science counts using only three numbers: 1, 2, ∞.

Throughout the social sciences we have fallen into this odd numeracy. We assume agents that are either hyperrational and informed or completely myopic. We consider models where time is instantaneous and place nonexistent. We represent our agents by a single prototype or have a world filled with so much heterogeneity that it results in unstructured noise. Our agents are either left isolated or are completely connected to one another. We seek an equilibrium in a system fraught with change.

Even with these restrictions, great progress has been made in understanding the complexity inherent in social systems. That being said, we now find ourselves with a new set of tools that, at least given our first decade of experience, appear poised to free us from the extremes and allow us to explore the interest in between.

The introduction of noise into a model provides another example of the interest in between. In noiseless systems, agents quickly get stuck, often in inferior configurations. With a lot of noise in the system, chaos reigns, and little, other than frenetic movement, is possible. Adding just a bit of noise to a system, however, often induces order and leads it toward optimization.

Social Complexity

Social agents fundamentally alter the behavior of complex systems. As previously discussed, when a car is about to crash, the interacting system of molecules in the bumper behaves very differently than the interacting system of passengers in the cabin, and it is only at the moment of impact that the behavior of these two types of agents (unfortunately) converges.

Thoughtful social agents can fundamentally alter the behavior of complex systems. In our Forest Fire model, we saw how, under homogeneous adaptation, the agents moved the system to a state that was at once optimal and fragile. When we allow these agents even more flexibility, they alter the system and allow it to enter new domains of behavior where the previously known "laws of nature" appear to no longer hold sway. In such a world, the discovery by the agents of microstructures like fire walls alters the physics of the world, and the agents are able to move the system into more interesting regions of behavior where they can get more out of the system while simultaneously reducing brittleness.

The simple choices of thoughtful agents can often have large impacts on the behavior of complex systems. In the self-organized critical systems we studied, we saw how thoughtful agents have the ability to transform the fat-tailed-distribution characteristics of these systems into any possible distribution. Even agents motivated by the simplest goals take a system characterized by avalanches on every scale and turn it into one with a rather orderly and short-lived periodic cycle.

As we admit finer gradations of thought, our systems are suddenly transformed into worlds rich in possibilities. To take just one example, consider our models of adaptive agents that can strategically communicate with one another. Giving agents the ability to communicate allows an unfolding of behavioral repertoires that begins by allowing simple self-recognition and, once that arises, creates a vibrant ecosystem of conversations, mimics, and other behaviors, which ultimately allow the system to operate in a new realm of previously inaccessible activity.

Throughout this work we are finding that adaptation implies more structure, not less. There has been an implicit assumption that because adaptive systems exist in the messy in-between, that the resulting models will be mired in incomprehensible layers of detail. Alas, the opposite appears to be the case: the messy in-between results in a reduction of complex behavior.

That is not to say that the emergent structures from messy models are always *easy* to predict or comprehend. We see a great deal of emergent perversity, as macrobehavior often differs radically from micromotives. Clearly there are worlds where the details do not matter, but we are still far away from having a theory of when this is so.

The link between the micro and macro is not as clear as we once thought. We must explore a new realm that both acknowledges the microfoundations of macrobehavior while simultaneously recognizing the potential for seemingly magical transformations that link one level to another. Of course, such magic is the impetus for the scientific exploration that in time will eventually lead to understanding.

The Faraway Nearby

Over the past decade, scholars have made much progress on understanding the complexity that surrounds us. The stimulus for this progress is a recognition that there are deep commonalities—ones that do not respect the usual academic boundaries—across the various systems we observe in the world. These commonalities drive the quest, and through a combination of new ideas and tools, we are slowly starting to reveal complexity's secrets.

The explorations over the past decade have allowed us to enter new realms of understanding. Alas, we are only now at the point where we have an innate sense that we have embarked upon an important mission. There are intriguing hints everywhere about the promise of the journey to come, and for the moment the path ahead is clear enough to proceed. No doubt that many obstacles will arise as we move forward, and while most of these will be easily surmounted, some of them may force us to alter our path in unanticipated directions. Nonetheless, the underlying journey seems sound, and great advances lack detailed maps.

An Open Agenda for Complex Adaptive Social Systems

> It then occurred to me that this was not the first time I had
> been given a map which failed to show many things I could
> see right in front of my eyes. All through school and
> university I had been given maps of life and knowledge on
> which there was hardly a trace of many of the things that I
> most cared about and that seemed to me to be of the greatest
> possible importance to the conduct of my life. I remembered
> that for many years my perplexity had been complete; and no
> interpreter had come along to help me. It remained complete
> until I ceased to suspect the sanity of my perceptions and
> began, instead, to suspect the soundness of my maps.
> —E. F. Schumacher, A Guide for the Perplexed

PART OF THE JOY OF SCIENCE is in chasing rainbows. Just when it seems as though the elusive quarry is within your grasp, it slips away and the quest must begin anew. With each new discovery about the nature of complex adaptive social systems, more questions arise.

In this appendix we outline some key elements of an open research agenda for the area of complex adaptive social systems, as well as some of the key contributions to date. These elements span a broad spectrum of science, and our hope is that they will provide a scaffolding around which various research efforts can take hold. Though far from complete, the items discussed demonstrate how, from a handful of provocative models, key insights into a more general science have emerged as complex systems research has moved from the fringe to the frontier.

A.1 WHITHER COMPLEXITY

In *A New Kind of Science* (2002), Steven Wolfram exhaustively investigates simple cellular automata. Wolfram makes several contributions, not the least of which is reminding us that simple rules in stark environments can generate complex aggregate behaviors. Thus, one explanation of why we see complexity is that simple rules generate it. This is an important idea, as we might have expected that simple rules in simple environments create only simple things.

Wolfram goes on to suggest that his restriction to simple models is an important one. In particular, he finds a natural bound on the computation a system can perform, and once this bound is reached, there is a universal equivalence that can allow one system to emulate another, *as long as* we are free to manipulate the inputs and reinterpret the outputs. Moreover, the suggestion is made that most systems (including some of the very simple ones he explores) have achieved this equivalence milestone.

In this sense, a *suitably set up and interpreted* pool table is the same as a human brain is the same as a weather system is the same as an economy. Of course, representations do matter, and what can be easily computed in one system is often difficult (but still possible) to compute in another. Thus, to use a simple cellular automata to calculate, say, the first few prime numbers may require a very complex calculation on how to set the initial conditions and interpret the output. This latter calculation could be much more "difficult" to compute than the original problem, just as the complexity of a compiler can far exceed the complexity of the programs it produces.

It is not clear what to make of computational equivalence. Yes, there is a threshold in which systems are related to one another, but given the difficulty of moving among them, is this anymore useful than saying that skateboards and Ferraris are equivalent means of moving about? Certainly, in some contexts, skateboards and Ferraris do share a lot in common, and knowing about one may lead to insights about the other. Yet, at some point, such insights break down.

Moreover, adaptation is not an inert force in these systems. In Wolfram's world, you either have behavior that is obviously simple or behavior that is (more likely than not) equivalently complex. The presence of adaptive agents begins to muddy this picture. Suppose a system composed of adaptive agents leads to a simple outcome. Such an outcome may present an opportunity for agents to further adapt and, in so doing, alter the outcome.

For example, imagine a stock market in which the prices oscillate with a known periodicity. Adaptive agents will respond to such a situation by selling when the prices are high and buying when the prices are low, altering the periodicity. The presence of adaptive agents drives the simple system into the complex regime where predictions, if at all possible, become extremely difficult.

A.2 What Does It Take for a System to Exhibit Complex Behavior?

One premise of complex adaptive social systems research is that more is different—that is, interacting agent systems take on a behavior that is qualitatively different from the behavior of any individual agent. While there are many examples of such behavior, there have not been systematic analyses of the fundamental conditions needed before a system will display such a response.

What is it about interacting agents that leads to complex behavior? Is there some behavioral threshold that must be breached before complexity can arise? Are there certain features that will prevent (or, for that matter, enhance) complexity? We know that certain simple rules can generate complex behavior, but so can complex rules. We also know that complex rules can generate simple behavior; for example, in general equilibrium theory sophisticated rules lead to stasis.

It appears that certain features of a system are more likely to lead to complex outcomes. For example, heterogeneity, adaptation, local interactions, feedback, and externalities all seem to induce more interesting patterns. Consider a market of homogeneous, hyperrational consumers trading goods at a single location. If these goods have no externalities among them, we would expect an equilibrium to arise. Now, introduce externalities, place these agents in space, or assign them different learning rules, and the market outcome is likely to exhibit even more complex patterns.

A.3 Is There an Objective Basis for Recognizing Emergence and Complexity?

Emergence is often subjectively identified during ex post analyses of systems. Is there an objective basis for defining emergence? If you put a frog in a blender and turn it on, there is only a macabre interest in the resulting chemical soup. If, however, you start with a chemical soup and run the blender backward, and out of the froth pops a fully formed frog, then something rather different has happened. Is there some easy and reliable way to separate out these two experiences? Of course, this would matter little if we weren't seeing so many frogs popping out of the froth of both nature and our models.

We can often explain emergent phenomena piece by piece. For example, patterns like a glider in the Game of Life can be understood by looking at each of its parts and the associated rules. Even for something as simple as a glider, however, this requires an enormous amount of work and rarely produces insight.

The path of the glider can be predicted without resorting to the microlevel rules. Thus, in a well-defined statistical sense, it requires less information to predict the path of the glider by thinking of it as a "thing" than it does to look at the underlying parts. In this sense, the glider has emerged (Crutchfield, 1994).

Note that this is not saying that emergence occurs when the parts cancel one another out. To the contrary, the parts are aggregating in complex and interesting ways. What they create, the emergent phenomena, has a statistical signature of its own, one that can be predicted more parsimoniously without looking at all of the parts. The concept of emergence has thus made the transition from a metaphor to a measure, from something that could only be identified by ocular magic to something that can be captured using standard statistics.

Identifying complexity has also been problematic. How do we separate complex systems from merely complicated ones? A check of the literature reveals that there are a variety of ways to measure complexity per se, but at this point in time there is no simple way to unify these definitions.

Trying to unify these definitions may be asking too much. Complexity can occur at many levels, including time, space, and interactions. Perhaps we are expecting too much if we want a single measure of complexity that captures all of our intuitions. Indeed, not having a uniform definition of, say, architectural beauty has not held back architecture, nor should the lack of a single definition of complexity hold back the science of complex systems.

A.4 Is There a Mathematics of Complex Adaptive Social Systems?

While computational models of complex adaptive social systems are a valuable theoretical tool, there may be other complementary tools that can be developed. The calculus allowed us to take certain, difficult-to-solve, nonlinear equations and reform them into simple linear problems. Is there a mathematics of complex adaptive social systems that will provide a similar transformation? Any simulation can be written as an instantiation of a recursive function, suggesting that a given model run is nothing more than a sequence of interconnected algebraic equations. But can we say something more general here?

Ultimately, we are seeking a simple explanation for complex behavior. While there are examples from cellular automata that suggest that the only way to predict the future behavior of the system is to let it fully run out, the obvious hope is that there are other opportunities to uncover more compact descriptions of complex behavior.

A.5 WHAT MECHANISMS EXIST FOR TUNING THE PERFORMANCE OF COMPLEX SYSTEMS?

While predicting and understanding complex adaptive social systems is a key goal of this research agenda, we would also like to have means for influencing the outcome of these systems, or as Karl Marx (1845) said, "philosophers have only interpreted the world in various ways; the point, however, is to change it." Much of the quest for good theory in this area has been driven by the desire to use it to improve the outcomes of real social systems. Indeed, "theories" of complex adaptive social systems are tested on massive scales every day, when governments implement various policies that often involve substantial resources and ultimately have tremendous impacts on the lives of countless citizens.

Insofar as real social systems behave according to the laws of complex adaptive social systems, what policies can be used to direct the outcomes of these systems? We know that some mechanisms, such as agent sorting in Tiebout worlds (Kollman, Miller, and Page, 1997) and double-auction markets, can naturally lead systems toward superior outcomes. What other mechanisms can be employed on this front?

Decentralized adaptive systems often display the ability to work around damage and self-repair—as Crichton's fictional Ian Malcolm said, "Life finds a way." Given this, what kinds of policies can lead to real change? We know that some complex systems enter regimes where they are very sensitive to the existing conditions, as illustrated by the well-known "butterfly effect" in chaos theory. Can such conditions be identified by observing the system, and if so, can they serve as useful leverage points from which to alter a system's subsequent behavior dramatically with only a small amount of effort?

A.6 DO PRODUCTIVE COMPLEX SYSTEMS HAVE UNUSUAL PROPERTIES?

What does it take for a complex system to undertake productive behavior? Previously we discussed the concept of the edge of chaos, whereby systems poised at the "Goldilocks point" can perform computation because the system is neither too static nor too chaotic. While metaphorically rich, is there a more exact statement of such a proposition?

What features of a complex system will either allow or prevent it from undertaking productive behavior? Moreover, is there some innate force that drives adaptive systems to this place?

A.7 Do Social Systems Become More Complex over Time?

Is there a time's arrow of motion on social systems that inevitably leads them to become more complex? If so, what drives such an arrow? We often see social artifacts, be they technological goods or institutions, becoming more complex over time. For example, software tends to become more resource intensive, feature laden, and sophisticated with time; razors acquire more blades; tax codes become more elaborate; laws more convoluted; and so on. Of course, there are movements toward simplicity on occasion, as tax codes are reformed, simplified software is developed (or rediscovered), and so forth.

In a world of thoughtful, interacting agents, complexity might emerge as those agents begin to "game" the system and, eventually, each other. There may be inherent forces in systems that drive out predictability. For example, in stock markets agents have incentives to find, and exploit, any regularities. In these types of systems, the actions of the agents result in the destruction of the regularity, and an increase in complexity.

However, there are other types of social systems in which the agents seek regularity. For example, institutions like legal or voting systems require a degree of predictability if they are to maintain legitimacy. As a system becomes more complex, causality is more difficult to infer and agents may actively attempt to reduce the apparent complexity by, for example, decoupling parts of the system or altering their behavior.

A.8 What Makes a System Robust?

Inherent in many complex adaptive social systems is a degree of robustness. For example, consider a city. Over very long time spans a particular city can remain whole, despite a multitude of changes in terms of both its fundamental structures and occupants. Similarly, a beehive can persist as an entity for many, many years even though its population is continually turning over (worker bees, constituting the vast majority of the hive, live for around one month in the summer and three months in the winter).

An analysis of robustness can take place at many levels. First, we can focus on the robustness of an entity relative to agent details. That is, which details of the agents matter in terms of maintaining the system's coherence? Second, we can consider the robustness of the entity to perturbations in the environment. What does it take for a system to persist in the face of external changes? Alternatively, we could frame the question as uncovering the factors that make a given system brittle.

A.9 CAUSALITY IN COMPLEX SYSTEMS?

Recognizing and understanding causality is one of the big challenges for agents within, and modelers of, complex systems. When agents in a complex system act, they may change the local or even the global path of events. Predictive agents, the kind that we often consider in social systems, need to have an understanding of the causal implications of their actions. Of course, we know that within the model is a full description of the dynamics, and hence the causes, of all actions, but such a description is unsatisfactory, as useful notions of causality require a much more compact and easily understood framework.

Modeling how agents simplify complexity so that they can predict and act is an important topic. There are a variety of techniques whereby apparent complexity can be condensed into useful approximations (indeed, the process of modeling itself is one such technique). Moreover, understanding how such simplifications themselves can influence the resultant complexity is also of interest.

A.10 WHEN DOES COEVOLUTION WORK?

There are nice examples, both natural and artificial, where coevolution allows the system to achieve previously inaccessible ends. For example, in biology we see "arms races" in which two species coevolve into very specialized niches, such as when a wren develops the ability to recognize and eject a cuckoo chick that has taken over its nest, and in response the cuckoo chick develops a finer begging call for food. In societies, we see coevolution between, say, legal systems and criminal activity.

Harnessing coevolution is a powerful way to adapt systems. Evolution needs to exploit opportunities inherent in the underlying structure of the world. Unfortunately, for most problems, the space of good structures is much smaller than the space of bad ones. In such a world, most of the feedback takes the form of "bad idea" rather than "good idea," and thus it is difficult for an evolutionary system to gain enough purchase to make rapid progress. Coevolution, however, lowers the fitness bar initially and, in so doing, allows systems to evolve more rapidly toward good structures. Over time, as the coevolving structures get better, the bar is raised and more selective pressure is exerted in the system. Coevolution allows the evolving entities to "challenge" each other progressively. By slowly, and automatically, ramping up the challenges, rapid evolutionary progress becomes possible.

A.11 When Does Updating Matter?

There are a variety of ways to activate agents in models of complex adaptive social systems, including simultaneously and a multitude of asynchronous options. As Bernardo Huberman has suggested, apart from people performing in marching bands or participating in military parades, most human updating is asynchronous. People update at different rates for various reasons ranging from their incentives to their access to information. Along with human systems, many other phenomena probably have asynchronous elements.

Does updating matter?

There are some theoretical results where agent timing can make a big difference in the outcome. For example, consider game theory. Games in which players sequentially take turns often result in outcomes very different from those in which the players move simultaneously.

We know of some complex systems, for example, Abelian sand piles, where the order of agent action makes little difference. Unfortunately, there are other situations in which updating does matter. The most dramatic example of this is in the Game of Life: in its synchronous instantiation it creates beautiful patterns and structures capable of universal computation, but under asynchronous updating the world quickly becomes barren.

The transition between synchronous and asynchronous updating is, in some ways, rather subtle as it only requires agents taking actions (or observing the actions of others) with just a slight offset from one another. Yet, as can be seen in the case of the Game of Life, it can lead to a dramatic difference—the difference between universal computation and mush.

A.12 When Does Heterogeneity Matter?

How do systems behave as we move from homogeneous to heterogeneous collections of agents? Much of empirical social science eliminates the impact of heterogeneity by relying on means of variables and thereby allowing the heterogeneity to cancel itself out. However, heterogeneity may not always cancel itself out, especially in the presence of dynamic feedbacks and other interconnections among the agents. Thus, in many models the tail (of the distribution) wags the dog. For example, the probability of a riot depends more on the number of people who are really angry than on the average level of discontent. To assume homogeneity is often to assume away much of what is interesting about the world.

There also may be thresholds as we increase the heterogeneity in a system that cause transitions into new forms of behavior. For example, the introduction of just a bit of heterogeneity into a homogeneous world may substantially alter the system's behavior. Moreover, as we further increase the amount of heterogeneity in the system, we might see yet another transition as the additional heterogeneity begins to homogenize the system. Perhaps it is the case that, as we increase heterogeneity, we move from simple systems to complicated ones back to simple ones.

A.13 How Sophisticated Must Agents Be Before They Are Interesting?

What happens to systems as we move from simple particle-like agents to sophisticated social ones? The first challenge here is to have a clear definition of the levels of agent sophistication. At one extreme, it is relatively easy to characterize a hyperrational, hyperinformed, hyperable agent that optimizes its behavior, given some objective. It is more difficult to define how one smoothly degrades such behavior and ends up at the opposite extreme of a myopic simpleton. Regardless of how we do this task—there may be many options—some interesting questions emerge.

Agent sophistication may not even be well defined. One cannot smoothly track behavior from optimizing to myopic. Intelligence, sophistication, whatever we want to call it, is more than one-dimensional and may not be easily "dialed in." However, within any one class of rules, often there does exist a dial. We can make the rules within a given class more "intelligent" by, say, varying the number of past periods remembered or future periods forecasted. We might also be able to move between classes of rules where one class embodies more intelligence than the other.

One key question is when does agent sophistication matter at all? We may have social systems where even a "dumb" agent can do "smart" things (think Forrest Gump). Another question is how rapidly does the system change as we degrade agent sophistication? Is there a smooth transition as we slowly degrade sophistication or are there rapid phase transitions where small changes in agent sophistication lead to large changes in system behavior. Of particular interest is how much difference does the very first movement away from full sophistication make?

The impact of agent sophistication is linked to the environment. For example, in a double-auction market the introduction of some irrationality is not likely to be noticed. In a game in which agents are trying to guess the guesses of other agents, however, a small amount of irrationality can wreak havoc. Not surprisingly, as cognitive feedback increases, that is, when it matters what you think I think you think and so on, the introduction of irrationality has more impact.

A related issue is whether we can easily calibrate computational agents to human ones. There are many potential uses for such a computational proxy. For example, by using artificial-adaptive agents on a class of games, we can identify a priori which games are likely to result in interesting experimental outcomes with human subjects. Another application of proxy agents would be in the design of new institutional forms. Here we could use proxy agents to explore large classes of potential institutions to help us identify which designs are most likely to meet our objectives.

A.14 WHAT ARE THE EQUIVALENCE CLASSES OF ADAPTIVE BEHAVIOR?

A full science of complex adaptive social systems requires a theory of adaptive behavior. Unfortunately, while there is typically one way to be optimal in the world, there are lots of ways to be adaptive. Given the potential multitude of adaptive behaviors, we may find ourselves confronting a zoofull of exotic agents, in which formulating a cohesive theory of adaptive behavior will be difficult, if not impossible.

Of course, it could be the case that, notwithstanding the vast array of potential adaptive behaviors, the behavior we observe in the real world is only a small subset of these possibilities. Alas, this seems like wishful thinking, as there do not seem to be any obvious convergence points, given the vast array of options.

Alternatively, it may be the case that a lot of adaptive behavior falls into a single equivalence class. If such a class exists, then the exact details of adaptation no longer matter much to the outcome. Thus, we can "let a thousand flowers bloom, a hundred schools of thought contend," and still be able to build a cohesive theory of adaptive agents.

Indeed, there already is some evidence of equivalence classes for adaptive behavior. Genetic algorithms define a fairly broad class of adaptive computation techniques that require potential solutions to be represented in a framework that allows "genetic" modifications linked to reproduction by performance. Implementations of such algorithms often differ in a variety of ways; for example, they may use very different kinds of representations, selection mechanisms, and choices of specific operators. Notwithstanding these choices, the resulting algorithms tend to perform more or less identically.

This experience with genetic algorithms suggests that certain types of equivalences are possible. The deeper suspicion is that such equivalences extend to a broad variety of other adaptive systems. To understand adaptation fully, we may first need to develop a well-defined taxonomy of adaptive behavior—that is, what are the key features that make a system adaptive. Once these features are defined, we may be able to extract their individual impacts from the specific context and ultimately find that, say, any system that embodies reproduction biased by performance will tend to imply a particular type of behavior.

There is also evidence from computational-learning models that different kinds of adaptation lead to similar predictions. In many games, Hebbian learning, replicator dynamics, fictitious play, and best-response learning lead to qualitatively similar individual and social behavior. That is not to say that these learning rules always lead to similar behavior, only that in many environments they do, and thus the details may not always be that important.

A.15 WHEN DOES ADAPTATION LEAD TO OPTIMIZATION AND EQUILIBRIUM?

While there is no imperative for adaptive systems to result in optimal structures, there are likely to be conditions under which simple adaptive systems produce optimal solutions. Under what set of general conditions will an adaptive system uncover the optimal solution to a problem? An alternative way to structure this question is to find the conditions under which certain problems become hard for adaptive systems to solve.

Although there are mathematical results, such as the No Free Lunch theorem (Wolpert and Macready, 1997), that suggest that no adaptive algorithm is uniformly best on all optimization problems, social agents likely exist in a fairly narrow part of the space of all problems. In some areas of adaptive algorithms, such as genetic algorithms, there have been attempts at defining "hard" problems, that is, problems with inherent structural elements that confound the adaptive mechanism.

Knowledge of when adaptation leads to optimization and or equilibrium would be useful in a variety of ways. With this knowledge, we would be able to deploy better our theoretical tools; for example, we would have some guidance as to when optimization techniques are likely to result in good predictions about the world.

Knowing more about adaptation would also help us refine our predictions in other contexts. For example, even though first-price, open-outcry auctions lead to the identical theoretical outcome as second-price, sealed-bid auctions, social systems may employ the former more than the latter because agents may have a far easier time adapting good strategies in such an environment ("bid a little higher if you can still make money" versus "if you reveal your true value and you have the second highest bid, then ..."). Practically, such knowledge might give us good insights into how to restructure situations so that adaptive agents can achieve better performance.

Neoclassical economists see systems primarily as either in equilibrium, heading toward equilibrium, or moving along a sequence of equilibria. Biologists and ecologists see systems as complex, dynamic networks of interactions. To say that economic phenomena are static and ecosystem phenomena are complex likely reflects field-specific modeling imperatives rather than deeper realities. Both types of systems can exhibit stasis and complexity.

A.16 How Important Is Communication to Complex Adaptive Social Systems?

Social systems are about interacting agents. Agents interact by taking direct actions that alter one another's worlds and indirect actions, via communication, that set the stage for the future. For example, two agents in a market may undertake substantial communication in the form of making offers, chatting one another up, and so on, prior to agreeing on the final terms and executing the trade. Much of current social science theory tends to ignore the communication phase of social interaction.

Communication among agents can have a profound effect on the behavior of a complex system. The ability to communicate expands the behavioral repertoire of the agents, introducing a variety of new opportunities ranging from the creation of abstract agreements to the opportunity to mimic and deceive. Communication can radically alter the performance of a social system; for example, ants leave pheromone trails that allow the colony to self-organize into a coherent mass for more efficient hunting. Of course, communication can also introduce detrimental outcomes by, say, allowing agents to collude in an auction. For example, in a 1996 Federal Communications Commission air-waves auction, bidders used the insignificant digits of their bid amounts to communicate their territorial intentions to the other participants. This form of communication appears to have arisen in earlier auctions in a more innocuous variation where it was used to display corporate logos and even the phone number of the congressman responsible for the auctions.

A.17 How Do Decentralized Markets Equilibrate?

It is time for the invisible hand to become more visible.[1] Although the theory of supply and demand makes fine predictions, it offers little in the way of helping us understand the mechanisms underlying these systems. As Hayek (1945, 530) suggested, there is a need to "show how a solution is produced by the interactions of people each of whom possesses only partial knowledge." Is there a coherent, plausible model that can help us understand the mechanism by which prices form in decentralized markets?

Uncovering such a mechanism will give us new insights into a well-known and long-standing issue in the social sciences. Moreover, it may prove to be a gateway into investigations of new market mechanisms that will have many practical applications. Improved communication and information-processing capabilities have opened up new possibilities for the creation of new markets. Similarly, government policies, such as selling the commercial rights to use the radio spectrum, have also necessitated the development of heretofore unknown auctions, like the "electronic simultaneous multiple-round" auction used by the Federal Communications Commission. Having an accurate theory by which to understand, design, and evaluate existing and potential markets would be of great value.

Finally, a solid theory of market mechanism might help explain why we see certain persistent market types. Around 2,500 years ago, simple auction rules began to emerge in Babylon. Since that time, societies have implemented an astonishing array of market mechanisms (although even these represent a small subset of all potential markets). Of these markets, only a handful are used to trade the vast majority of the world's goods and services. Is there some driving force that causes such a convergence? How much do particular mechanisms, or parts of mechanisms, actually matter?

A.18 When Do Organizations Arise?

Organization is a fundamental issue spanning the world of complex systems: atoms form molecules, molecules form cells, cells form organisms,

---

[1] Indeed, you have to wonder about any theory that relies on "invisible" driving forces.

organisms form firms, and firms form nations. What underlies this apparent order? When will organizations emerge or dissipate?

The question of organization touches on many of the previous themes. Organizations can emerge from a variety of substrates. They tend to persist, even though their constituent parts do not. They exist in a state that allows them to be productive without being dissipative.

A.19 What Are the Origins of Social Life?

The origin of life question has played a central role in the biological sciences. Alas, the origin of social life has had much less attention. Such questions lie at the heart of understanding our world. How do we recognize social life? What are the minimal requirements for it to arise? What are the deep, common elements in social systems that transcend time and agents? Is social life inevitable?

Various research efforts in complex systems have shown how key social features, like cooperation or communication, can emerge. Yet, even these models tend to rely on some previously defined atomic structures. For example, agents are assumed to have strategic frameworks or are endowed with the ability to send and receive communication tokens. Is it possible to unwind these models further, allowing even more to emerge? Ultimately, can we realize *in silico* a dream similar to Darwin's, where starting from so simple a beginning we see endless social forms, most beautiful and most wonderful, arise?

Practices for Computational Modeling

> Yet the question of its *modus operandi* is still undetermined.
> Nothing has been written on this topic which can be
> considered as decisive—and accordingly we find every where
> men of mechanical genius, of great general acuteness, and
> discriminative understanding, who make no scruple in
> pronouncing the Automaton a pure machine, unconnected
> with human agency in its movements, and consequently,
> beyond all comparison, the most astonishing of the
> inventions of mankind. And such it would undoubtedly be,
> were they right in their supposition.
> —*Edgar Allan Poe, Southern Literary Messenger*

In 1769 Baron Wolfgang von Kempelen created an "automaton" chess
player. The device consisted of an artificial "Turk" seated behind a
cabinet full of mechanical marvels. The Turk would move the pieces
with its mechanical arm and even nod its head disapprovingly when
the opponent made an illegal move (apparently a feature that was well
utilized when it played Napoleon Bonaparte). The automaton was a
sensation around the world.

The most important feature of the automaton was indeed its clever
mechanical design—especially the feature that allowed a skilled human
chess player to remain concealed to the audience. While consciously
hidden assistants are not a scientific concern, the area of computational
modeling is new enough that adhering to a set of best practices will
do much to advance the acceptance and productivity of this approach.
Like all scientific fields, the biggest danger with computational models
is not outright fraud (a relatively rare event across all fields of science),
but the unconscious acceptance of faulty results. It is easy for scientists
to fool themselves, and all scientists must be their own harshest critics
and do everything they can to maintain and expose the full integrity of
their work.

In this appendix, we begin to outline some practices that should help
promote quality science in computational modeling. Note that most of
these practices have very little to do with computational methods per
se, and that is by design—good computational modeling is more about
modeling than computation.

B.1 Keep the Model Simple

Making sure that your model has just enough of the right elements and no more is the most fundamental practice for any kind of modeling, and computational work is no exception. With a few well-placed lines of a pen, an artist can represent a complex world in a very simple and understandable way. Scientific modelers must aim for a similar level of simplicity and clarity. Modeling is like stone carving: the art is in removing what you do not need.

It is often easy to recognize a simple, well-formulated model after the fact, as such models have a strong intuitive appeal. Getting to this point usually requires a combination of skill, practice, effort, revision, and art—a mix of abilities that is difficult to teach directly. A first step in developing such skills is appreciating the elegance of seminal models from a variety of fields.[1] Great artists study the masters; so too must great modelers.

Once simplicity is achieved in the model, many of the practices suggested here become much easier to attain.

There is a necessary tension between pursuing simplicity and exploiting the ability of computational models to interconnect key parts of the world. For example, to understand the dynamics of an epidemic, we need to understand how diseases are transmitted between people; how people come in contact with one another via transportation, work, and home environments; and how people react once they become ill or hear of illness in others. As we incorporate each of these elements into the model, it becomes more complicated—though avoiding this type of complication is difficult if we truly want to understand this type of problem. We can admit such complications into our models if we are careful to keep each of the constituent parts simple. Good models strip phenomena down to their essentials, yet retain sufficient complication to produce the needed insights.

B.2 Focus on the Science, Not the Computer

The most important feature of a computational model is the model, not the computer. Thus, computational models must be justified solely by the model and not on their clever (or fast, current, etc.) algorithm (or, for

---

[1] In economics, works like Akerlof (1970), Hotelling (1929), Schelling (1978), and Stigler (1961) offer a few examples.

that matter, hardware, software, etc.).[2] New technological developments may enhance our ability to explore better existing models or create new ones, but without a solid model underlying the work, such improvements are meaningless. Thus, being able to take a simple model that produces high-quality science and scaling it up by, say, adding more agents or performing some additional experimental conditions may be a good application of new technological developments. Using lovely real-time graphical displays to illustrate your results can also be productive, as long as there is good science underlying the results.

B.3 The Old Computer Test

One way to promote both of these practices is to write a computational model as if it must run on a very slow and limited computer. Writing a program under such a constraint will often force the modeler to simplify the work in productive ways. (Of course, once the model is finalized, you can always run it on the best-quality machine available.) Some of the best examples of "computational" models, such as Schelling's Segregation model, do not even require a computer.

B.4 Avoid Black Boxes

Each part of a model must be as clear and accessible as possible. To achieve this end, modelers should always default to simple, straight-forward mechanisms for each element and avoid having parts of the computation rely on "black boxes" that are composed of massive amounts of code rife with assumptions and choices that are essentially hidden from any potential consumers of the model. When models implement such code, it is difficult to determine whether the outcome that is observed is being driven by some quirk hidden within the black box or by a more fundamental law of nature.[3]

Of course, what is a black box to one person might be a fundamental component to someone else. For example, one approach pursued in the cognitive science and artificial-intelligence communities is the creation of relatively large and complicated computational models of cognitive processes. These models are composed of various components like

[2]Indeed, it is not clear to us why speakers feel a need to discuss issues surrounding hardware or software choices (or, even worse, display computer code) during scientific seminars. Such discussions are best left for other contexts.

[3]The fundamental issue here is captured in a cartoon by Sidney Harris that has two scientists discussing sets of equations on a blackboard divided by the words "then a miracle occurs." One says to the other, "I think you should be more explicit here in step two."

memory, categorization, inference, and problem solving, each of which is implemented through an often complicated set of assumptions and specializations for a given problem domain. If we accept these efforts as well-formulated and understood models of adaptive behavior, then one could use them as the basis for creating social systems of adaptive agents, notwithstanding the black box nature of the computation involved. Nonetheless, wherever possible, computational modelers should follow the dictates of Ockham's razor and favor the adoption of simple structures. If a simple, low-level adaptive algorithm is able to capture the behavior of a system adequately, then it should be preferred over more complicated mechanisms.

B.5 Nest Your Models

There is often an opportunity to create models where special cases of the assumptions result in well-known (and hopefully understood) examples. Nesting standard models within computational ones is usually a very natural process, as computational models are often employed to extend standard models in interesting, but previously inaccessible, directions. Once nested, it is easy to compare the model's predictions in the special cases with known results, and then to show how the model verifies known results and observations or, if not, to explain why there is a divergence.

In the course of developing and fully understanding a computational model, there are usually plenty of opportunities to include "sanity checks," that is, special conditions under which the behavior of the model is known a priori. Almost any component of a computational model can be turned off and replaced by a simple alternative. For example, if the model relies on adaptive agents, complicated objective functions can be temporarily replaced with simple ones to demonstrate that the agents can find the optima in such a case. These types of experiments are a good way to check the basic foundations of the model and should be a routine part of creating any computational model.

B.6 Have Tunable Dials

One way to nest models is to rely on "tunable" dials for controlling key assumptions. One interesting dial for economic problems would be a way to tune agent rationality. Ideally, this dial would allow us to take a fully optimizing agent and slowly degrade its behavior toward less and less rationality. Such a dial would provide fodder for a variety of interesting research questions: Under what conditions does the dial matter? Does the behavior of the system undergo slow changes as we turn the dial, or does the system experience rapid phase transitions where small movements of the dial result in abrupt changes in the system? What effect does the very first movement of the dial away from optimality have on the system?

Finding a simple way to represent a rationality dial would be a great advance, and at the moment even the existence of such a dial is an open question. That being said, there are areas where dials can be easily employed; for example, we can vary the degree of look-ahead used by an agent, the amount of processing, the number of interactions, and so on.

B.7 CONSTRUCT FLEXIBLE FRAMEWORKS

Often it is possible to create computational models with simple, flexible frameworks that "get filled in" by the computation. For example, genetic algorithms rely on representations of potential solutions that are fairly general, and it is the evolution of the system that fills in the details. A well-designed framework puts very few a priori constraints on the model, and thus the outcome is rich in possibilities. Such frameworks provide enough flexibility so that the model can explore areas that were not fully anticipated by the researcher. It is not uncommon for, say, a genetic algorithm to yield an answer that initially appears wrong, only to find on closer examination that the algorithm discovered a perfectly sensible, but wholly unanticipated, result.

Note that even the simplest of frameworks can result in outcomes that are difficult to understand. For example, consider a simulated neural network composed of a group of artificial neurons controlled by a straightforward adaptive algorithm. Such a network provides a very flexible framework that is constructed of extremely simple components. However, understanding the outcome of this system can be very difficult, as it produces a nonlinear function that is often difficult to unravel. Thus, in addition to simplicity and flexibility, we need models that are transparent in their operation and outcomes.

B.8 CREATE MULTIPLE IMPLEMENTATIONS

Even with an apparently well-defined model, there may be a variety of choices that must be made before it can be fully implemented. Most of the time, such choices matter little to the outcome of the model, but on occasion they may lead to important differences (and hence insights). Therefore, it is always useful to implement key modeling choices in a variety of ways.

One useful way to facilitate the creation of multiple implementations of a model is to pose it in relatively general terms to different modelers and have each of them separately define an implementation. By comparing these various renditions, one can begin to identify the key issues that need to be addressed. An alternative technique is to have at least two groups separately code the model (preferably using two different computer languages). Not only does this process help clarify the important issues, but it also results in two versions of the model that can be run in parallel to confirm results and gain insights (Axtell et al., 1996).

Even within a single, well-defined computer program, there are often opportunities to implement key features of a model in different ways. For example, assumptions about, say, probability distributions, agent matching, timing, and so on, can be manipulated easily. Computational modelers should always try to have multiple implementations of any features of the model that will be closely associated with scientific claims. Thus, it is often useful to implement a variety of adaptive algorithms to make sure that any particular claims of the impact of adaptive behavior are not tied to some quirk in a specific algorithm.

B.9 Check the Parameters

Along with implementing various parts of the model in different ways, computational models should always be subject to a sensitivity analysis of key parameters. This can either be done manually or through the use of automated techniques, like Active Nonlinear Tests (ANTs) (Miller, 1998). Under ANTs, researchers specify their main assumptions (including parameter values, distribution choices, and so on) and then use an automated procedure that is designed to "break" the model by searching over acceptable variations of the assumptions. ANTs can be used to identify a model's inherent limits and key driving assumptions.

B.10 Document Code

Most of the coding effort devoted to a computational model, like most programming, is devoted to documenting the model and getting data into and out of the system, as opposed to specifying core behavior. Care and time spent in this domain are necessary for ensuring that the results can be fully analyzed and easily tied to the exact conditions that produced the outcome. Software tools, such as Concurrent Versions Systems (CVS), can assist in the tracking of the various revisions made to the program during a research project.

B.11 Know the Source of Random Numbers

Models that rely on random numbers must ensure that the pseudo-random-number generators used are adequate to the task. It is easy for the uninitiated to make serious errors in this realm; for example, using only the lower-ordered bits coming from a pseudo-random-number generator can be problematic as in some algorithms these bits alternate in value. Users should always be aware of the algorithmic details of their generators and take whatever appropriate actions (like shuffling the resulting values) are needed to avoid problems.

B.12 Beware of Debugging Bias

In all scientific work, there are natural human biases that often confound good practice. All scientists have some expectations (and dreams) about the results they will find, and when these are met, it is natural to accept the findings. If you look at the historical estimates of, say, the speed of light, you will find that they do not randomly vary centered on the currently accepted value, but rather they converge on this value biased by previously published estimates (Henrion and Fischhoff, 1986).

A similar bias can happen in computational models with respect to debugging. When modelers observe results that are not as expected, they are likely to spend a lot of effort debugging their code. When their expectations are met, little such effort is expended. Thus there is an inherent tendency in researchers that could, if sufficient caution is not exercised, bias results toward prior expectations.

B.13 Write Good Code

Various software development practices can improve the code underlying a computational model. Some key lessons from software development that are applicable here include the following. First, there is a trade-off among scope (the number and types of features included in the software), quality (the ability of the code to be understood, modified, and executed without error), cost (both in terms of development and maintenance), and delivery date. The trade-off is such that improving one of these areas causes one of the other areas to degrade (or, as it is sometimes stated: scope, quality, cost, time, choose three). Second, the majority of time on a well-run software development project is not spent in coding, but rather it is devoted to developing a solid, well-articulated design for the software. The third important lesson from software development is that the cost of correcting errors dramatically increases with the amount of time that elapses before the error is identified and fixed. Thus, an error in the design stage costs ten times more to correct in the coding stage and a hundred times more to fix after the program is in use. Finally, there are dramatic differences in the practices of professional versus amateur software coders (even among the professional coders, the productivity can differ by a factor of ten or more), and all computational modelers should make sure that their skills are sufficiently advanced for the task at hand. McConnell (1993) provides a nice overview of the basics for writing high-quality, extendable, easily communicated code.

Good software development is extremely difficult, and given the complexity of the underlying task, newness of the area, and rapid changes in technology, it is not too surprising that ideas about how best to accomplish this task are in constant flux. Much like business management fads, every few years new methods of software development are proposed that promise to solve all of the existing problems.

While it is doubtful that such a silver bullet will emerge anytime soon, there may be parts of certain development methodologies that are useful for computational modelers. For example, the Extreme Programming movement incorporates a number of practices that closely match the needs of scientific modelers, six of which we list here: key features of the program are added in the order of most value; overall planning is limited to current needs; each component of the program is associated with a comprehensive unit test that ensures that it is implemented appropriately; an effort is made to improve the code (refactor) whenever possible; development time estimates are based on recent experience; and new versions of the program are released on very short time scales.

Scientific researchers undertaking the development of ambitious software projects should consider adopting a real software development "process" (versus the usual "code and fix" methodology). Although real processes do introduce some additional overhead costs into the project (especially initially), the higher-quality software that is produced by such methods allows these overhead costs to be quickly recovered.

B.14 Avoid False Precision

When reporting computational work, be aware that there is an appropriate level of precision that can be associated with the results. This concept holds at all levels of the model. At the lowest level, make sure that numeric results are appropriately reported. At a higher level, given the inherent nature of all modeling, beware of making too much of subtle differences. The important results from a model are typically not very subtle and tend to be obvious, both qualitatively and quantitatively, across a variety of conditions.

B.15 DISTRIBUTE YOUR CODE

Computational models should be easily accessible to other researchers. Models that follow the practices outlined here are often more accessible to others, because they entail simple, high-quality code, that is well written and documented, and thus easily understood and implemented by others. Code for published models should be easily available to others so that they can replicate the results. Some researchers are also willing to distribute code prior to publication (perhaps with some stipulations on its use).

B.16 KEEP A LAB NOTEBOOK

Once developed, computational models are often deployed in a way that is very similar to laboratory experiments in biology or chemistry. In these areas, there is a long tradition of keeping a laboratory notebook that details particular hypotheses, the experiments designed to test the hypotheses, and the outcomes and conclusions of these tests. Such notebooks not only provide an important historical record of the work, but they may also give the researcher additional insights and understanding about the system. Such research notebooks may have a similar value for computational experiments.

B.17 PROVE YOUR RESULTS

Whenever possible, computational results should be clarified and verified as thoroughly as possible. Computational results should always be placed in their appropriate statistical context. More important, researchers should strive to eliminate as many alternative hypotheses as possible through well-designed experiments. Indeed, a better name for this section might be "disprove your results"—researchers should always try to disprove their key results to the best of their ability.

Researchers should try to use alternative means to solidify key findings. Thus, it is often possible to apply more traditional tools to parts of the analysis by, say, proving a theorem about a particular phenomena observed in the model. Although such techniques may not work on the whole model, they often work on particular parts or special cases. Whenever possible, the analysis of computational models should be enhanced with complementary modeling efforts.

B.18 REWARD THE RIGHT THINGS

Like any branch of science, one needs to reward the right accomplishments. While it may be true that lovely graphics, advanced coding techniques, frontier hardware, and so on may enhance computational models, ultimately it is the resulting science that must be judged. Scientific judgments in this area should focus not on the computer per se, but on the quality and simplicity of the model, the cleverness of the experimental designs, and the new insights gained by the effort.

Bibliography

Abbott, Edwin A. 1884. *Flatland: A Romance of Many Dimensions*. New York: Dover, 1952.

Akerlof, George A. 1970. "The Market for 'Lemons': Quality Uncertainty and the Market Mechanism." *Quarterly Journal of Economics* 84:488–500.

Albin, Peter S. 1975. *The Analysis of Complex Socioeconomic Systems*. Lexington, Mass.: Lexington Books.

Anderson, Phil W. 1972. "More Is Different." *Science* 177:393–96.

Andreoni, James, and John H. Miller. 1991. "Can Evolutionary Dynamics Explain Free Riding in Experiments?" *Economics Letters* 36:9–15.

———. 1995. "Auctions with Adaptive Artificial Agents." *Journal of Games and Economic Behavior* 10:39–64.

Arrow, Kenneth J. 1951. *Social Choice and Individual Values*. New Haven: Yale University Press.

Arthur, W. Brian. 1994. "Inductive Reasoning and Bounded Rationality." *American Economic Review Papers and Proceedings* 84:406–11.

Arthur, W. Brian, John H. Holland, Blake LeBaron, Richard G. Palmer, and Paul J. Tayler. 1997. "Asset Pricing under Endogenous Expectations in an Artificial Stock Market." In *The Economy as an Evolving Complex System II*, edited by W. Brian Arthur, Steven Durlauf, and David Lane, 15–44 Redwood City, Calif.: Addison-Wesley.

Axelrod, Robert. 1984. *The Evolution of Cooperation*. New York: Basic Books.

Axtell, Robert. 2001. "Zipf Distribution of U.S. Firm Sizes." *Science* 293:1818–20.

Axtell, Robert, Robert Axelrod, Josh Epstein, and Michael Cohen. 1996. "Aligning Simulation Models: A Case Study and Results." *Computational and Mathematical Organization Theory* 1:123–41.

Bak, Per. 1996. *How Nature Works*. New York: Springer-Verlag.

Bednar, Jenna, and Scott E. Page. 2006. "Can Game(s) Theory Explain Culture? The Emergence of Cultural Behavior within Multiple Games." *Rationality and Society* 18:345–73.

Camerer, Colin. 2003. *Behavioral Game Theory: Experiments in Strategic Interaction*. Princeton, N.J.: Princeton University Press.

Camerer, Colin, and Teck-Hua Ho. 1999. "Experience-Weighted Attraction Learning in Normal Form Games." *Econometrica* 67:837–74.

Challet, D., and Y. C. Zhang. 1997. "Emergence of Cooperation and Organization in an Evolutionary Game." *PhysicaA* 246:407–18.

———. 1998. "On the Minority Game: Analytical and Numerical Studies." *PhysicaA* 256:514–32.

Coates, Robert M. 1956. "The Law." In *The World of Mathematics*, Vol. 4, edited by James R. Newman, 2268–71. New York: Simon and Schuster.

Crutchfield, James P. 1994. "The Calculi of Emergence: Computation, Dynamics, and Induction." *PhysicaD* 75:11–54.

Darwin, Charles. 1859. *The Origin of Species*. New York: Mentor, 1958.

Dawkins, Richard. 1976. *The Selfish Gene*. Oxford: Oxford University Press.

Dennett, Daniel. 1995. *Darwin's Dangerous Idea: Evolution and the Meanings of Life*. New York: Touchstone.

Eliot, Thomas S. 1942. "Little Gidding." *New English Weekly*.

Epstein, Joshua M. 1999. "Agent-Based Computational Models and Generative Social Science." *Complexity* 4:41–60.

Epstein, Joshua M., and Robert Axtell. 1996. *Growing Artificial Societies*. Cambridge, Mass.: MIT Press.

Fischer, Shira H. 2004. "Bee Cool." *Science Now* 624:3.

Frette, Vidar, Kim Christensen, Anders Malthe-Sorenssen, Jens Feder, Rorstein Jossang, and Paul Meakin. 1996. "Avalanche Dynamics in a Pile of Rice." *Nature* 379:49–52.

Friedman, Milton. 1953. *Essays in Positive Economics*. Chicago: University of Chicago Press.

Gardner, Martin. 1970. "Mathematical Games: The Fantastic Combinations of John Conway's New Solitaire Game." *Scientific American* 223:120–23.

Glaeser, Edward L., Bruce Sacerdote, and Jose A. Scheinkman. 1996. "Crime and Social Interactions." *Quarterly Journal of Economics* 111:507–48.

Gode, Dan K., and Shyam Sunder. 1993. "Allocative Efficiency of Markets with Zero Intelligence Traders: Market as a Partial Substitute for Individual Rationality." *Journal of Political Economy* 101:119–37.

Grannoveter, Mark. 1978. "Threshold Models of Collective Behavior." *American Journal of Sociology* 83:1420–23.

Hayek, Friedrich. 1945. "The Use of Knowledge in Society." *American Economic Review* 35:519–30.

Hebb, Donald O. 1949. *The Organization of Behavior: A Neuropschological Theory*. New York: Wiley.

Henrion, Max, and Baruch Fischhoff. 1986. "Assessing Uncertainty in Physical Constants." *American Journal of Physics* 54:791–98.

Hillis, W. Daniel. 1990. "Co-evolving Parasites Improve Simulated Evolution as an Optimization Procedure." *PhysicaD* 42:228–34.

Holland, John H. 1975. *Adaptation in Natural and Artificial Systems*. Ann Arbor: University of Michigan Press.

Holland, John H., Keith J. Holyoak, Richard E. Nisbett, and Paul R. Thagard. 1986. *Introduction: Processes of Inference, Learning, and Discovery*. Cambridge, Mass.: MIT Press.

Holland, John H., and John H. Miller. 1991. "Artificial Adaptive Agents in Economic Theory." *American Economic Review, Papers and Proceedings* 81:365–70.

Hooke, Robert. 1665. *Micrographia*. Reproduced digital edition, Oakland, Calif.: Octavo.

Hopfield, John J. 1982. "Neural Networks and Physical Systems with Emergent Collective Computational Abilities." *Proceedings of the National Academy of Sciences* 79:2554–58.

Hotelling, Harold. 1929. "Stability in Competition." *Economic Journal* 39:41–57.

Huberman, Bernardo, and Natalie Glance. 1993. "Evolutionary Games and Computer Simulations." *Proceedings of the National Academy of Sciences USA* 90:7716–18.

Ishii, Hiroshi, Scott E. Page, and Niniane Wang. 1999. "A Day at the Beach: Human Agents Self-Organizing on the Sand Pile." *Advances in Complex Systems* 2:37–63.

Jacobs, Jane. 1984. *Cities and the Wealth of Nations*. New York: Random House.

Jervis, Robert. 1997. *System Effects: Complexity in Political and Social Life*. Princeton, N.J.: Princeton University Press.

Judd, Kenneth. 1997. "Computational Economics and Economic Theory: Complements or Substitutes?" *Journal of Economic Dynamics and Control* 21:907–42.

———. 1998. *Numerical Methods in Economics*. Cambridge, Mass.: MIT Press.

Kirkpatrick, S., C. D. Gelatt Jr., and M. P. Vecchi. 1983. "Optimization by Simulated Annealing." *Science* 220:671–80.

Kirman, Alan. 1997. "The Economy as an Interactive System." In *The Economy as an Evolving Complex System II*, edited by W. Brian Arthur, Steven Durlauf, and David Lane, 491–531. Redwood City, Calif.: Addison-Wesley.

Kollman, Ken, John H. Miller, and Scott E. Page. 1992. "Adaptive Parties in Spatial Elections." *American Political Science Review* 86:929–37.

———. 1997. "Political Institutions and Sorting in a Tiebout Model." *American Economic Review* 87:977–92.

———. 1997. "Landscape Formation in a Spatial Voting Model." *Economic Letters* 55:121–30.

———. 1998. "Political Parties and Electoral Landscapes." *British Journal of Political Science* 28:139–58.

———. 1998. "Computational Political Economy." In *The Economy as an Evolving Complex System II*, edited by W. Brian Arthur, Steven Durlauf, and David Lane, 461–90. Redwood City, Calif.: Addison-Wesley.

Kreps, David M., Paul Milgrom, John Roberts, and Robert Wilson. 1982. "Rational Cooperation in the Finitely Repeated Prisoners' Dilemma." *Journal of Economic Theory* 27:245–52.

Krugman, Paul. 1995. *Development, Geography, and Economic Theory*. Cambridge, Mass.: MIT Press.

Kuan, Chung-Ming, and Halbert White. 1994. "Adaptive Learning with Nonlinear Dynamics Driven by Dependent Processes." *Econometrica* 62:1087–114.

Langton, Christopher G. 1989. "Artificial Life." In *Artificial Life*, edited by Christopher G. Langton, 1–47. Redwood City, Calif.: Addison-Wesley.

———. 1990. "Computation at the Edge of Chaos: Phase Transitions and Emergent Computation." *PhysicaD* 42:12–37.

Leady, James. 2006. "Strategy Sophistication, Decision Costs, and Experience in Multiple Game Environments." In *Essays on Bounded Rationality and Strategic Behavior in Experimental and Computational Economics*. Ph.D. dissertation, University of Michigan.

LeBaron, Blake. 2001. "Stochastic Volatility as a Simple Generator of Apparent Financial Power Laws and Long Memory." *Quantitative Finance* 1:621–31.

Ledyard, John. 1986. "The Scope of the Hypothesis of Bayesian Equilibrium." *Journal of Economic Theory* 39:59–82.

Le Guin, Ursula K. 1969. *The Left Hand of Darkness*. New York: Walker.

Lewis, John P., and Richard C. Webb. 1997. *The World Bank: Its First Half Century* (Vols. I and II). Washington, D.C.: Brookings Institution Press.

Li, Wentian, and Norman Packard. 1990. "The Structure of the Elementary Cellular Automata Rule Space." *Complex Systems* 4:281–97.

Lohmann, Susanne. 1993. "A Signaling Model of Information and Manipulative Political Action." *American Political Science Review* 88:319–33.

Malthus, Thomas. 1798. *An Essay on the Principle of Population, as It Affects the Future Improvement of Society with Remarks on the Speculations of Mr. Godwin, M. Condorcet, and Other Writers*. London: J. Johnson.

Marx, Karl. 1845. *Theses on Feuerback (Thesis XI)*. First published as an appendix to Engel's *Ludwig Feuerback and the End of Classical German Philosophy* (1886).

McBeath, Michael K., Dennis M. Shaffer, and Mary K. Kaiser. 1995. "How Baseball Outfielders Determine Where to Run to Catch Fly Balls." *Science* 268:569–73.

McConnell, Steve. 1993. *Code Complete: A Practical Handbook of Software Construction*. Redmond, Wash.: Microsoft Press.

McKelvey, Richard, and Thomas Palfrey. 1998. "Quantal Response Equilibria for Extensive Form Games." *Experimental Economics* 1:9–41.

Metropolis, N., A. Rosenbluth, M. Rosenbluth, A. Teller, and E. Teller. 1953. "Equation of State Calculations by Fast Computing Machines." *Journal of Chemical Physics* 21:1087–92.

Miller, John H. 1988. "The Evolution of Automata in the Repeated Prisoner's Dilemma." In *Two Essays on the Economics of Imperfect Information*, 49–97. Ph.D. dissertation, University of Michigan.

———. 1996. "The Coevolution of Automata in the Repeated Prisoner's Dilemma." *Journal of Economic Behavior and Organization* 29:87–112.

———. 1998. "Active Nonlinear Tests (ANTs) of Complex Simulations Models." *Management Science* 44:820–30.

Miller, John H., Carter Butts, and David Rode. 2002. "Communication and Cooperation." *Journal of Economic Behavior and Organization* 47:179–95.

Miller, John H., and Scott Moser. 2004. "Communication and Coordination." *Complexity* 9:31–40.

Miller, John H., and Scott Page. 2004. "The Standing Ovation Problem." *Complexity* 9:8–16.

Mitchell, Melanie. 1997. *Introduction to Genetic Algorithms*. Cambridge, Mass.: MIT Press.

Mitchell, Melanie, James P. Crutchfield, and Peter Hraber. 1994. "Dynamics, Computation and the 'Edge of Chaos': A Re-examination." In *Complexity: Metaphors, Models, and Reality*, edited by George Cowan, David Pines, and David Melzner, 497–513. Reading, Mass.: Addison-Wesley.

Moore, Christopher, and Martin Nilsson. 1999. "The Computational Complexity of Sandpiles." *Santa Fe Institute Working Paper.*

Newman, Mark J. 2003. "The Structure and Function of Complex Networks." *SIAM Review* 45:167–256.

Ostrom, Elinor. 2005. *Understanding Institutional Diversity.* Princeton, N.J.: Princeton University Press.

Ott, Edward. 1993. *Chaos in Dynamical Systems.* Cambridge: Cambridge University Press.

Packard, Norman H. 1988a. "Intrinsic Adaptation in a Simple Model for Evolution." In *Artificial Life,* edited by Christopher G. Langton, 141–55. Redwood City, Calif.: Addison-Wesley.

———. 1988b. "Adaptation toward the Edge of Chaos." In *Dynamic Patterns in Complex Systems,* edited by J. A. S. Kelso, A. J. Mandell, and M. F. Shlesinger, 293–301. Singapore: World Scientific.

Page, Scott E. 1997. "On Incentives and Updating in Agent Based Models." *Computational Economics* 10:67–87.

———. 1998. "On the Emergence of Cities." *Journal of Urban Economics* 45:184–208.

———. 2006. "Essay: Path Dependence." *Quarterly Journal of Political Science* 1:87–115.

Plott, Charles. 1967. "A Notion of Equilibrium and Its Possibility Under Majority Rule." *American Economic Review* 57:787–806.

Poe, Edgar Allan. 1836. "Maelzel's Chess-Player." *Southern Literary Messenger* 2:318–26.

Pynchon, Thomas. 1973. *Gravity's Rainbow.* New York: Viking.

Rapoport, Anatol, and Melvin Guyer. 1966. "A Taxonomy of 2 × 2 Games." *Yearbook of the Society of General Systems Research* 11:203–14.

Richardson, Lewis. 1960. *Statistics of Deadly Quarrels.* Pittsburgh, Pa.: Boxwood Press.

Rust, John, John H. Miller, and Richard Palmer. 1992. "Behavior of Trading Automata in a Computerized Double Auction Market." In *The Double Auction Market: Institutions, Theories, and Evidence,* edited by Daniel Friedman and John Rust, 155–98. Redwood City, Calif.: Addison-Wesley.

———. 1994. "Characterizing Effective Trading Strategies: Insights from a Computerized Double Auction Tournament." *Journal of Economic Dynamics and Control* 18:61–96.

Samuel, Arthur. 1959. "Some Studies in Machine Learning Using the Game of Checkers." *IBM Journal of Research and Development* 3:210–29.

Samuelson, Larry. 2001. "Analogies, Adaptations, and Anomalies." *Journal of Economic Theory* 97:320–67.

Samuelson, Paul. 1954. "The Pure Theory of Public Expenditure." *Review of Economics and Statistics* 36:387–89.

———. 1999. Foreword. In Michael Szenberg, *Passion and Craft: Economists at Work.* Ann Arbor: University of Michigan Press.

Schelling, Thomas. 1978. *Micromotives and Macrobehavior.* New York: Norton.

Schumacher, E. F. 1978. *A Guide for the Perplexed.* New York: HarperCollins.

Shannon, Claude. 1948. "A Mathematical Theory of Communication." *Bell System Technical Journal* 27:379–423 (July) and 623–56 (October).

Simon, Herbert. 1955. "On a Class of Skew Distribution Functions." *Biometrika* 42:425–40.

———. 1969. *The Sciences of the Artificial.* Cambridge, Mass.: MIT Press.

Smith, Adam. 1776. *An Inquiry into the Nature and Causes of the Wealth of Nations.* London. Edited by Andrew Skinner. New York: Pelican Books, 1970.

Stigler, George J. 1961. "The Economics of Information." *Journal of Political Economy* 69:213–25.

Stuttaford, Andrew. 2002. "Chick-Tac-Toe: A Prized Fight in Las Vegas." *National Review*, December 23.

Tesfatsion, Leigh. 1997. "How Economists Can Get Alife." In *The Economy as an Evolving Complex System II*, edited by W. Brian Arthur, Steven Durlauf, and David Lane, 533–64. Redwood City, Calif.: Addison-Wesley.

———. 2006. "Agent-Based Computational Economics: A Constructive Approach to Economic Theory." In *Handbook of Computational Economics*, Vol 2: *Agent-Based Computational Economics*, edited by Leigh Tesfatsion and Kenneth L. Judd, 831–80. Amsterdam: North-Holland.

Thompson, D'Arcy W. 1917. *On Growth and Form.* New York: Dover, 1992.

Tiebout, Charles M. 1956. "A Pure Theory of Local Expenditures." *Journal of Political Economy* 64:416–24.

Turing, Alan. 1937. "On Computable Numbers, with an Application to the Entscheidungsproblem." *Proceedings of the London Mathematical Society* (series 2) 42:230–65.

Ulam, Stanislaw. 1976. *Adventures of a Mathematician.* Berkeley: University of California Press.

Von Neumann, John, and Oskar Morgenstern. 1944. *Theory of Games and Economic Behavior.* Princeton, N. J.: Princeton University Press.

Vriend, Nick J. 2000. "An Illustration of the Essential Difference between Individual and Social Learning, and Its Consequences for Computational Analyses." *Journal of Economic Dynamics and Control* 24:1–19.

Watts, Duncan J. 2003. *Six Degrees: The Science of a Connected Age.* New York: Norton.

Weaver, Warren. 1958. "A Quarter Century in the Natural Sciences." In *The Rockefeller Foundation Annual Report*, 7–15. New York: Rockefeller Foundation.

Wellman, Michael, Amy Greenwald, Peter Stone, and Peter Wurman. 2003. "The 2001 Trading Agent Competition." *Electronic Markets* 13(1):4–12.

Wolfram, Stephen. 1984a. "Computer Software in Science and Mathematics." *Scientific American* 251:188–203.

———. 1984b. "Universality and Complexity in Cellular Automata." *PhysicaD* 10:1–35.

———. 2002. *A New Kind of Science.* Champaign, Ill.: Wolfram Media.

Wolpert, David H., and William G. Macready. 1997. "No Free Lunch Theorems for Optimization." *IEEE Transactions on Evolutionary Computation* 1:67–82.

Zambrano, Eduardo. 2004. "The Interplay between Analytics and Computation in the Study of Congestion Externalities: The Case of the El Farol Problem." *Journal of Public Economic Theory* 6:375–95.

Index